Consulting in a Book
The Ultimate Desk Reference
for Engineers and Scientists

Successful Consultant Training LLC

805-451-7658

Figures

Foreword

After several decades of consulting, I decided to codify the techniques, tools, models, ideas, concepts, and other tips that have helped me succeed. In addition, I included suggestions from many business and marketing experts that I've known over the years.

I formed Successful Consultant Training LLC as a vehicle to provide training to engineers and scientists. I began by offering courses online. Then I developed a series of webinars geared towards engineering consulting firms, the type of firms I worked for as a staff engineer to an Executive before branching out to solar energy development. The webinars reflected what I had learned throughout my career.

When thinking about how to market these webinars, I decided that writing a book about the webinars would be a good marketing tool. However, as I was writing the book, I realized that the book itself would be a novel way to train engineers and scientists. Not only is my material different than that currently being offered by other training firms, but I think that a training tool that can be used by an engineer while working is a more effective way to teach and learn. With this in mind, this book was born and given a subtitle "The Ultimate Desk Reference for Engineers and Scientists". The material is directed primarily at the private sector.

This approach to training is different. The sole proprietor keeps this book handy and consults it frequently for advice on business development, working with clients, and problem solving.

In a consulting firm, the operations manager and other senior personnel will find a number of ideas in the book directed towards operations issues (e.g. hiring practices, performance appraisal methods). I direct your attention primarily to Chapter 3 and Appendix A and B.

In addition, the book can be a useful training vehicle. At the least, each staff member would benefit by having a copy at their desk. But, managers and supervisors can take it a step further.

For example, supervisors or group leaders could organize periodic (weekly is best) "brown bag lunches" to discuss one chapter or even one topic in a chapter. They'd ask the attendees to read that chapter ahead of the lunch meeting. The group would discuss the chapter's content in the meeting, and conclude the meeting by discussing which technique or tool attendees would apply at the next opportunity. In a follow-up meeting, attendees could report back on how they have applied the new techniques to their work. Repeat this process for each new topic of discussion and action. The management involvement would greatly increase the effectivity of the training.

SCT provides a Manager's Guidebook for many of the topics in the book. Each Guidebook contains color slides of the material, suggested "workshop" ideas for training sessions, and a variety of templates and checklists that support the subject.

There are some major benefits to this approach to training:

- Deliberate and constant professional development is the very best way to raise confidence, competence, and performance levels;
- Using visual aids in training dramatically increases retention;
- Strong learning environments are characterized by exceptionally high staff engagement;
- Training saves money through reduced turnover, and increases revenue when techniques are successfully implemented and applied to operations;
- This type of training does not impact billability;
- The cost is minimal. Simply purchase the book and guidebook for each member of the team;
- Frequent training sessions give managers an opportunity to influence habits and instill the importance of learning new skills;
- Weekly training increases camaraderie and puts some peer pressure on staff to apply their new skills;
- Weekly training is by definition, repetitive. *Repetition is a key to learning.* This is where other training methods fall short.

The word "reference" is in the title of this book because the book contains hundreds of tools, techniques, and ideas in skill and knowledge areas that are important for all engineering consultants, engineers, scientists, engineering managers, operations managers, or other consultants. The idea is that each member of the staff can pick and choose which tools or techniques they want to learn and apply (i.e. cafeteria style). This learning freedom brings more engagement.

You will not find some of the topics in this book in typical training courses for consultants. For example, there is a chapter on marketing psychology, a very important subject that is often overlooked. There is also a chapter on problem solving and decision-making. Most consultants solve problems for their clients, but they haven't received training in these subjects. How consultants communicate with their prospects and clients is a critical part of selling. We devote a chapter to this topic, and introduce a truly unique method of improving communication and marketing skills.

For the individual trainee, think of this book as a very large checklist. When you begin a task, review the appropriate chapter in this book for ideas and tips. For example:

- Consult Chapter 2 to learn about:
  - The habits of successful consultants
  - Why a growth mindset sets you on a path to success
- Consult Chapter 3 to learn about:
  - The Compleat Business Development Model, a novel approach to business development
  - A systems approach to marketing, which offers many useful benefits
  - Developing effective Lead Generation Systems
  - The science of Fascination and how it can improve your communication and persuasion skills and provide management with helpful tools
  - How you can use your authentic communication style to influence and persuade your prospects and clients
  - How to "read" your client
  - How to create a strong value proposition how to apply techniques of marketing psychology to your business development activities

    - How to position your firm before meeting a new prospect

- Consult Chapter 4 to learn about:
    - Great questions to ask in your client feedback session
    - Holding a successful kick-off meeting
    - Probing a client's problem
    - Risk management strategies
    - How to run a successful client meeting

- Consult Chapter 5 to learn about:
    - The Problem-Solving Framework, an in-depth treatment for solving a client's problems
    - How to make effective decisions for your clients.

- Consult Appendix A to learn about:
    - The unique Proposal Success Equation
    - Creative storyboards for client presentations
    - Questions to ask for a Go No-Go determination
- Consult Appendix B to learn about:
    - The 2X Client Planning Strategy
    - Building client loyalty
    - Building deeper relationships with your clients

If you consult this book often, you will gradually acquire many new techniques that will enhance your performance, as well as the Operation's performance.

I. Four Key Success Skills for Consultants

The Importance of Increasing your Net Worth as Human Capital

Every consulting firm relies heavily upon the effectiveness of their "human capital", defined here as the productivity and creativity contributed by staff. Each of us has a human capital net worth to the firm we work for. As individuals, we strive to maximize our net worth to the firm. When we do this, we benefit as individuals, and the consulting firm benefits in terms of increased revenues and profits. In other words, your human capital net worth is an important source of competitive advantage for you personally, and for your operation or office. This book offers four ways to increase your human capital net worth as a consultant.

A few years ago, I was speaking to a room of about 80 consultants. I asked them a question: "What does a consultant need to do to make a contribution to the operation's success?" What I was really asking is *how do you increase your net worth?* Here are some responses I received:

- Facilitate client meetings
- Communicate effectively
- Be a solid team player/leader
- Effectively manage project budgets and schedules
- Have a growth mindset/ be a learner
- Maintain and grow a network
- Maintain a strong reputation
- Have a clear focus on client services
- Be a good project manager
- Make effective decisions on behalf of clients
- Exercise excellent marketing and sales skills
- Build rapport and trust with clients
- Seek and give performance feedback
- Demonstrate solid analytical skills
- Write winning proposals and presentations

I thought about this list for some time. Gradually, the pieces fell together. The items on this list appeared to fit four overreaching topics: business development, client services/service delivery, problem solving, and mindset. Furthermore, competence on these topics appeared to be consistent with my experience.

We conclude this section with a quote from Rafaella Sadun, Nicholas Bloom, and John Van Rennen that appeared in the September-October 2017 Harvard business article "Why Do We Undervalue Competent Management?" (Sadun, Bloom & Van Rennen, 2017)

> *"While to some extent the availability of skills is shaped by a firm's specific context, managers can play a critical role by recognizing the critical importance of employee's basic skills and providing internal training programs.*

The SCT Success Chain

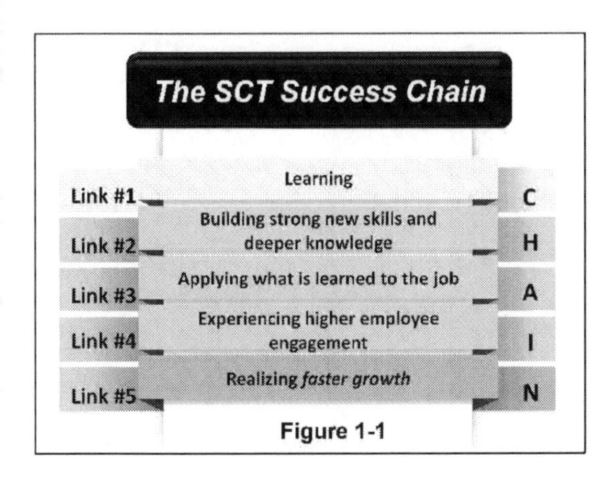

Figure 1-1

The SCT Success Chain is shown in Figure 1-1. I think these five steps, or links in a chain, in progression, are vital in order to achieve significant improvement and faster growth for individuals, operations or offices. These links are closely related to the human capital net worth of the employees.

The first link is "Learning". Everything begins with learning. *Additional learning is needed to advance to a higher level of achievement and results.* Clearly, operational success and innovation -- necessities today-- are dependent on learning. Dr. W. Edwards Deming, my mentor early in my career, said that success will come from continuous learning -- advice that I took to heart.

The second link is "b\Building strong new skills and deeper knowledge". The goal here is to strengthen skills and knowledge that are directly related to performance improvement and are congruent with the needs of the organization.

The third link is "Applying what is learned to the job". It is relatively easy to learn, but applying what you have learned is a much bigger hurdle that many find difficult to achieve. In a consulting firm (or other organization) this responsibility rests with both the staff and the operation or office management. *Of the five links, this one is far and away the most difficult.* Management must play an important role to help the staff apply what is learned.

The fourth link is "Experiencing higher employee engagement". When the staff learns new skills and knowledge and then applies it to the job, they will feel much more engaged and empowered. This is a powerful combination that creates a virtuous cycle! Daniel Pink, in *Drive: The Surprising Truth About What Motivates Us* (Pink, 2012), as well as Michael Mankins and Eric Garton in *Time Talent Energy* (Mankins & Garton, 2017) emphasize that skills mastery leads to stronger engagement and higher productivity.

The fifth link in the SCT Success Chain is "Realizing faster growth". With the previous four links in place, your operation will, without a doubt, benefit in terms of increased revenues and higher profits.

In summary, everything flows from learning, which increases the net worth of the staff and results in superior individual and operational performance.

The Value of Systematics

Systematics is the science of creating and optimizing systems and processes. The most important mentor in my career has been Dr. W. Edwards Deming, likely the most successful consultant in the last century if measured by results. Deming was a statistician, management consultant, and professor. I met him at New York University where he taught in the Quantitative Analysis group. Dr. Deming was a major contributor to sampling theory, now used in polling. In the 1950's he spent a lot of time in Japan, teaching quality control techniques to the auto industry. He was instrumental in Japan's turnaround from building poor cars to the highest quality cars in the world. He then consulted with the major auto makers in the U.S.

Dr. Deming emphasized the importance of understanding your work in terms of a system. He would say that if you didn't know how to express your work as a system, you didn't understand your work. One of his most well-known and profound quotes is "Your system is perfectly designed to give you the results you are getting." During my career, it became patently obvious that when you look at a task in terms of a system, rather than a collection of sub-tasks, many advantages become accessible. I became convinced that systems knowledge is an important, but not well recognized, pathway to higher net worth and greater success. We devote an entire chapter on this topic, applied to business development.

The Four Key Skills Consultants Must Have

Let's return to the four key skills for consultants. Raising the level of one's human capital net worth one of the best ways to increase one's value as a consultant. Decades of experience in consulting lead me to the conclusion that four key skills are the major contributors to net worth, or value. Having an expertise isn't enough. Engineers must have effective communication skills, display divergent thinking, have a growth mindset, have solid problem-solving abilities, and a strong focus on innovation. The four key skills address these needs. *Note that these skills cannot be outsourced.*

Performance coach Brendon Burchard conducted research on the habits of highly successful people. One of his findings indicated a close relationship between mastering the basic skills needed on a job and performance. He also mentioned that job performance drives engagement in his book, *High Performance Habits: How Extraordinary People Become That Way* (Burchard, 2017).

A Consultant's Mindset

The first skill is having a consultant's mindset. How you think about your work, your attitude, your confidence, your energy, and your belief system can have a substantial effect on your results. Your habits and how you approach your work will impact your level of success.

About twenty years ago, Dr. Carol Dweck, a professor at Stanford University, performed research on achievement and success, and discovered that our beliefs have an unusually strong impact on our ability to succeed in life. We'll address this issue in the next chapter.

In addition, my observations regarding staff performance of hundreds of staff members over the years indicate that a few habits tend to support success. Working on your "success habits" can have a real positive impact on your human capital net worth.

Business Development

The second skill is business development. This is a broad term that includes many skills, such as marketing, sales, writing proposals, making presentations and more. Embedded within these skills are sub-skills that are also essential for success in business development. Let's consider the following examples:

1. **Systems in Marketing.** Earlier, we discussed the importance of systems. My experience with systems in marketing shows just how powerful an approach this is for a consulting firm. There is no reason why every firm, and every individual, shouldn't be applying a systems approach to their marketing.
2. **Psychology of marketing.** Few engineering firms focus on the benefits of psychology in their marketing, despite the fact that keys to success in marketing are persuasion and influence. The reason engineering consulting firms avoid this aspect of marketing is that "emotional marketing" is a "fuzzy" concept to most engineers and scientists.
3. **Value proposition.** The value proposition is the kernel, or most important part, of a marketing strategy. The best value propositions appeal to a client's strongest, most impactful decision-making drivers. An effective value proposition requires a significant effort to determine the client's core needs, their interpretation of value, and how to customize an offer that is congruent with the client's definition of value.
4. **Communication skills**. Experts in marketing will tell you that a key ingredient for success is differentiation. You must stand out in some way to grab the attention of the client or prospect. You must influence them to buy from you. Perhaps our most effective differentiator is our personality, or how we communicate with others. However, few people in marketing realize this or take advantage of this. Everyone has a unique way of communicating, but most people do not take advantage of that uniqueness to provide their highest distinct value. We explore this fascinating topic.

A considerable portion of this book is devoted to business development. Marketing and sales don't seem to come naturally for engineers and scientists, so putting a significant effort in this area pays substantial dividends.

Client Service Strategies

The third key skill for consultants is client service strategies, or how we deliver services to clients. In other words, how well we understand our clients and how we interact with our clients is critical to our success. Building loyalty is a key to that success. However, in my experience, the most important aspects of client interaction and communication are the questions we ask and the feedback we request. Many consultants do not realize that getting client feedback on their performance can be the best marketing tool in their quiver.

Problem Solving

The fourth key skill for consultants is problem solving. Virtually every consultant's primary role is to solve their client's problems. Yet, most consultants haven't received training in problem solving.

Problem solving isn't just about solving problems. You must consider all possible solutions to a problem and weigh the consequences of each in order to choose the solution that results in the best possible outcome to the problem. Our chapter on problem solving and decision-making covers what we call the "Nine-Step Problem Solving Framework," a detailed process for problem solving that will increase anyone's problem solving prowess.

So, there you have it: the four key skills for consultants, especially engineering consultants or technical consultants. Consultants who have a "growth mindset" will continually work to increase their skills in these four areas. That is a perfect lead-in to our next chapter. But first, a quick note on how this book is organized.

Organization of this Book

The book is organized around the four key skills, which are further discussed in Chapters 2 through 5.

Two appendices are included that cover additional business development topics that do not fit directly into the Compleat Business Development model discussed in Chapter 3.

Appendix A is called Preparing Successful Proposals and Presentations. It offers a new approach to crafting proposals and presentations.

Appendix B is called Marketing Strategies for Key Clients. This section presents a number of strategies for developing better relationships and winning more work from your key clients.

II. Key Skill No. 1: The Consultant's Success Mindset

This chapter deals with our mindset, beliefs, and habits. These factors can have a significant impact on our ability to succeed, either as a booster or as a hurdle.

What is a Mindset?

A mindset is a collection of beliefs or attitudes. I like Dan Kennedy's definition of mindset:

"Mindset is about what stone you have set your mind in."

Not surprisingly, our beliefs affect our performance. Dr. Carol Dweck of Stanford University conducted research into the effects of different mindsets on success in *Mindset: The New Psychology of Success* (Dweck, 2008). She defined two broad types of mindsets: a fixed mindset and a growth mindset. The following table illustrates the differences between these mindsets.

Fixed Mindset	Growth Mindset
Effort is devalued	Effort leads to mastery
Low level of patience	Success takes time
Avoid challenges	Willing to "risk it"
Other's success isn't good	Success is inspirational
Cannot accept criticism	Use criticism to improve
Quick to give up	Always curious
Intelligence or talent is a fixed trait: have it or don't	Learning is a path to better results
A world of scarcity (only so many pieces of the pie)	A world of abundance (the pie gets bigger)

The table clearly demonstrates the ways in which a growth mindset lends itself to consultant success, whereas a fixed mindset can hinder it

As you will see in this book, I like to express concepts in terms of an equation. Boiling down Dr. Dweck's theory of mindset, I get the following:

Growth Mindset = Desire + Goals + Efficacy + Focus – Ego

People with growth mindsets typically have a desire to take on challenges, set goals for motivation, engage the subconscious, believe that they can meet the goals (efficacy), and have a strong desire to bring clarity to what they want to accomplish. People with a growth mindset do not allow their ego to get in the way of progress. By decoupling our ego from our beliefs, we open ourselves to further growth. We will revisit these terms throughout the book.

When I find myself struggling to tap into my growth mindset, I try to remember to look at a sign on my wall that says, "If things didn't go my way today, there are only two reasons why: a) I didn't give it the time necessary, or b) I didn't know enough."

The Consultant's Cycle of Success

Throughout my career as a consultant, in which I supervised hundreds of staff members, a few factors stood out that closely related to performance. For example, staff who knew where they wanted to go had at least a general strategy for getting there. High performers tended to be excited or enthusiastic about the work they were doing, and they weren't often distracted from their quest. Also, successful employees sought out more experienced people who could help them where needed.

I codified these seven behaviors that tend to be evident in successful consultants into what I call the Consultant's Cycle of Success, shown in the Figure 2-1.

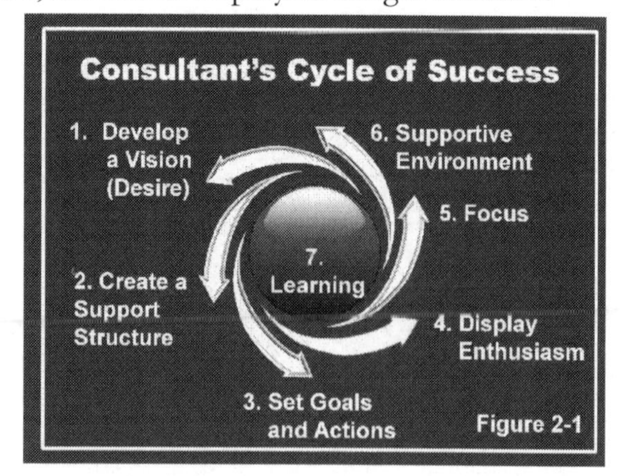

Figure 2-1

Everything Begins with a Vision

Here is what Napoleon Hill said about having a vision:

> *"The psychological effect of having a personal vision is to impress the vision upon your subconscious mind so strongly that it accepts that purpose as a pattern or blueprint that will eventually dominate your actions."*

In order to expand the results in your business, it's critical that you have a compelling vision of your future. If your view of that future is not significantly better than where you are now, there is no reason -- and no motivation -- to change anything.

What do we really mean by the term "vision"? In 2009, my partner and I, with help from a private equity firm, founded a solar energy company named Pacific Valley LLC. Our vision was to "become a leader in utility-scale distributed solar energy development in central California". Was this a good vision?

First, a vision is a high level, aspirational statement of a long-term objective or direction. Our vision sounds reasonable based upon that definition.

Second, you must be excited about this vision; in other words, you should be emotionally attached to it. My partner and I put everything we had into the company. We essentially dropped all other commitments and focused on this opportunity alone.

Third, it must be realistic. You must believe that you can reach the vision. At the time, there was little information on the efficacy of this undertaking because the utility-scale solar plant solution using photovoltaic panels was in its infancy. However, we did have some prior experience with other solar technologies and a couple large rooftop solar projects, so we had confidence that we could succeed, even though we were a very small firm.

Fourth, the vision must be clear. There should be no question about what it is, and a pathway to achieving it should be evident. Our vision was a clear statement of our objective.

Finally, a vision statement should serve as your "why". It is the reason why you do what you do, it is why you are energized about the future, and the reason why you work towards more specific goals that will provide a visible path to the vision.

The Absolute Power of Goal Setting

The next step in the Cycle of Success is setting goals. For many, this is a tough one. Let's begin with a couple quotes. The first one, by comedian Lily Tomlin, is both humorous and insightful:

"I always wanted to be someone. I guess I should have been more specific."

Second, a quote by legendary philosopher, speaker, and coach, Jim Rohn:

"The real value in setting goals is not in their achievement. Rather, the major reason for setting goals is to compel you to become the person it takes to achieve them."

I think that goals help you assemble, organize, clarify and prioritize your thoughts, intention, decisions, and actions. A goal creates intention, the precursor to creative activity. Intention also brings desire, which leads to action.

The concept of goals and their consequences has been with us for a long, long time. Here is a quote from an ancient Vedic Upanishads text from 700 B.C.:

"You are what your deepest desire is.
As is your desire, so is your intention.
As is your intention, so is your will.
As is your will, so is your deed.
As is your deed, so is your destiny."

Note that your deepest desire is your vision. Your intention is a trigger for action. Your will is your commitment to act. Your deed is your specific goal. And, your destiny is your vision.

Goals must be crafted in such a way that aids in their completion. My experience indicates that goals should have these seven characteristics:

1. Goals must be specific and measurable, so that progress and accomplishment can be recognized;
2. Goals must be realistic, that is, achievable, and you must believe that you can achieve the goal;
3. There must be a timeframe, or schedule for goal completion;
4. You must believe that you can accomplish the goal. Self-belief = self-efficacy;
5. The responsible party must be indicated;
6. There must be sufficient resources available to complete the goal;
7. The benefit of achieving the goal must be clear. Goals must provide a benefit that creates a drive, or motivation, to accomplish the goal. In other words, *there has to be a "why"*.

Continuing with the example of my vision for my own company, the top-level goal we used at Pacific Valley to get started was:

"Secure permits and electric transmission agreements for ten project sites in Fresno and Madera Counties, and obtain power purchase agreements for five of these projects."

In order to achieve this top-level goal, we created the following six lower level goals:

- Search for electric substations with sufficient capacity and low upgrade costs
- Search for retired farmland near these substations
- Engage a realty team with personal knowledge of the landowners
- Negotiate land leases or purchases
- Initiate a public relations effort
- Secure permits to build

Then, each of these goals was broken down even further. For example, the public relations effort's primary benefit was reducing possible opposition:

- Engage a public relations firm (Week 1-2)
- Create a company website (Weeks 4-8)
- Craft messaging and prepare brochures (Weeks 4-8)
- Conduct opinion surveys (Weeks 6-10)
- Meet with elected officials (Weeks 8-10)
- Donate to elected officials' re-election campaigns (Week 9)
- Join key organizations (i.e., the Economic Development Corporation) (Week 9)
- Meet with key organizations (i.e., the Farm Bureau) (Week 10)

I began this section saying that goals are tough. I say this from experience. There seems to be an unending list of reasons why people do not see the benefit of goal setting. Among those are:

- Your colleagues don't use goals, so why should you?
- You have a fear of failure, or you tried them once and they weren't achieved;
- You just don't believe that goals work;
- You don't realize how powerful goals can be;
- You don't have a vision, so there is no motivation to use goals;
- And, perhaps the toughest one of all: setting and working towards goals takes real time and effort.

That is why most consultants don't set goals, or they do so with a minimum level of commitment. If your belief in goals is wavering, re-read the Vedic Upanishads quote.

If you are not setting goals and reviewing them often, try it for a few weeks. You will quickly realize the benefits!

Finally, a message for managers and supervisors regarding goal attainment: Teresa Amabile and Steven Kramer, in their book *The Progress Principle*

(Amabile & Kramer, 2011) mention that when progress is made towards goals, employee engagement soars.

Creating a Support Structure for Success

The third step in the Cycle of Success is a personal support structure. Brendon Burchard's quote regarding structure address the importance of this concept:

"Structure infuses discipline and adds clarity."

When we say structure, we mean activities that help us stay organized and focused. Examples include agendas, lists (to do lists, checklists), daily planning, a coach or mentor, and mental triggers.

Daily planning is the first activity you do each morning (not reading your emails!). For example, use a template that lists the key actions you must complete each t day (including actions that move you towards your goals) and the contacts you need to call or email. The purpose of daily planning is to focus on what is important, not what is urgent. Recall this advice from Steven Covey author of the book *The Seven Habits of Highly Effective People* (Covey, 1993):

"In addition, daily planning helps you avoid major distractions that interrupt your progress during the workday. Make daily planning a habit."

A mental trigger, or shortcut, is any stimulus that reshapes our thoughts and/or actions. I call them "intention initiators". For example, before I meet with a client or prospective client, I have an intention initiator that says, "It is my job to fire up my client with enthusiasm". This always gets my discussion off to a good start! Another example is my daily planner worksheet, which I fill out at the start of the day. It has a statement at the top of the form that says, "What am I going to be enthusiastic about today?" This initiator reminds me to start the day with enthusiasm. I have a sign at my desk that says "Email" with a skull and crossbones under it to remind me that emails are other people's agendas, not mine. Bottom line, create a few mental shortcuts that trigger desired behaviors through the day.

Of course, having a coach is the most effective "structure". A coach will ensure that you keep your eye on what matters. In the absence of a coach, the structure we discussed here can make a difference.

Get Enthusiasm!

The fourth step in the Cycle of Success is enthusiasm. Napoleon Hill in *The Laws of Success* (Hill, 2004) says this about enthusiasm:

> *"Enthusiasm is a state of mind that inspires and arouses one to put action into the task at hand."*

I often tell people that enthusiasm is an energizing force, like electricity. Without it, you will be like a dead battery! Enthusiasm is a motivating force that raises your level of interest, helps overcome procrastination, and promotes action.

Enthusiasm Creates DRIVE!
Figure 2-2

What does enthusiasm look like? For example, you wake up in the middle of the night with a new idea. Or, you just finished an 80-hour work week and you still feel energized. When you are enthusiastic, or in the flow, time passes quickly.

What You Focus on is Manifested

Focus is the fifth step in the Cycle of Success. Here's what Isaac Newton said about focus:

> *"If I ever made any valuable discoveries, it has been due more to focus than to any other talent."*

Focus includes deciding which actions you will *not* take. Focus is directly related to a feeling of being in control, seeing a pathway to goal achievement, more effective problem solving, and better decision-making. This brings us to pattern recognition through the Reticular Activating System (RAS). I first learned about the RAS at a seminar conducted by Chet Holmes, a well-known teacher of marketing principles, called *The Ultimate Sales Machine* (Holmes, 2007).

The Reticular Activation System is an efficient pattern recognition system in your brain that we don't use often enough. The RAS *increases* our focus on things that we think about. For example, if you read your goals every morning, the RAS will program your subconscious mind to work on these goals 24 hours a day. The reverse is also true. If you focus on the negative, your brain will give you more negative stuff, so be careful. A good use of the RAS is to help you solve a problem. Just before going to bed, your mind is less busy. This is the time to concentrate on the problem and engage your RAS. This works! Thomas Edison said, "never go to sleep without a request to your subconscious". He filed for 2,332 patents!

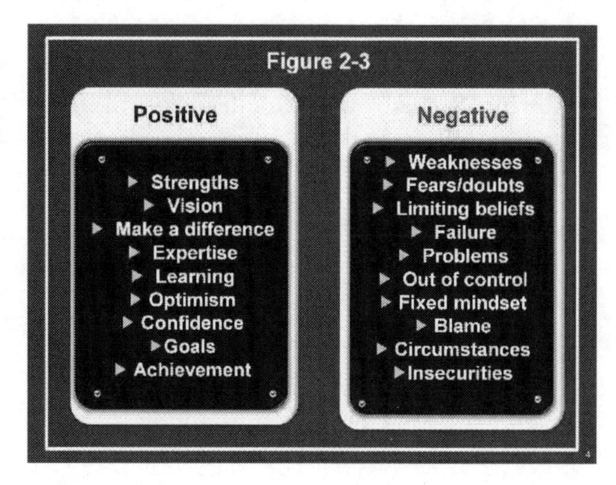

Finally, focus results in a sense of optimism and positivity. One concept of the field of positive psychology that appears quite useful for consultants is that we can influence what we focus on. Essentially, we can choose to focus on positive things or negative things. Examples of "things" are shown in Figure 2-3.

Psychologists say that things like limiting beliefs, fears, doubts, etc. are always with us. We can't eliminate them. It's part of being human. However, rather than trying to "work on them", the idea behind positive psychology is to focus more on the positives than the negatives. This is our choice. The idea here is to spend more time focusing on the left side of the table (the positive) as opposed to the right side (the negative). When we do this, we become more creative, we have more energy and enthusiasm, and we are happier. It follows that we become more successful.

Learning is a Mindset

The sixth factor in the Cycle of Success is learning. E.D. Hess, a Professor and author of "Learn or Die: Using Science to Build a Leading-Edge Learning Organization " (Hess, 2014) says this about learning:

> *"More than ever, organizations and individuals must either be continuously learning, adapting, and improving, or risk professional obsolescence."*

Learning is a mindset. I like to label consultants who have this mindset "Learning Professionals". A Learning Professional defines everything he or she knows *as conditional* – subject to change based on new evidence – and can help decouple egos from our beliefs. A Learning Professional has a healthy respect for the magnitude of what he or she doesn't know, and has a belief system that requires continuous learning.

Learning also applies to consulting firms. Here are five characteristics found in firms that put a lot of importance and effort in learning:

First, frequent feedback is provided, which supports high employee engagement. Feedback is provided at the task and assignment level, not only at an annual performance review. Group leaders, or supervisors should have less than ten direct reports. In this environment, the supervisor can spend the time needed to engage the staff on a personal basis.

Second, there should be permission for everyone to speak freely. This means staff can go to managers and discuss any topic without the possibility of retribution. Open communication in the office is the norm. People do not talk about staff "behind their back".

Third, professional development is strongly encouraged. When staff is learning, they are internally or intrinsically motivated. The responsibility for development is a shared responsibility between the employee and management. While the office can provide support in the form of travel to conferences, bringing in experts to talk on various topics, or providing mentoring, the employees devote their personal time as well.

Fourth, there is conditional permission to make mistakes provided that the mistakes are used as a way of learning from them. Unfortunately, we've been educated in a culture where mistakes are bad (i.e. high test scores are good, low test scores are *bad.)* As humans in a work environment, we all make mistakes. It is the responsibility of management to ensure that learning occurs when mistakes happen. If the staff is afraid to make a mistake, they will not take the risks needed to succeed. The staff will be reticent to offer new ideas.

Early in my career, I was managing a project to permit a waste coal power plant. We had a technical issue with obtaining air quality data, requiring a rapid expenditure of money.

Unfortunately, I did not notify the client about this out-of-scope work until we spent over $100,000 of unauthorized funds. The client took this opportunity to change consultants and get $100,000 of labor and equipment for free. My boss at the time saw this as a learning situation, which was not the reaction I anticipated, but this learning experience served me well throughout my career, as I am sure it did my colleagues who vowed on that day to never make the same mistake I had. The lesson I learned here was invaluable.

The final characteristic found in firms that put a strong emphasis on learning is the level of control staff has over the learning. When an individual has a say in what is to be learned, their motivation becomes more intrinsic (e.g. "I *want* to learn this") rather than extrinsic (e.g. "My boss says I *have to* learn this.").

Nothing Happens Without ACTION!

The seventh and last factor in the Cycle of Success is ACTION! Nothing happens in the absence of action. Here are three quotes that shine some light on this statement.

The following Chinese proverb expresses this thought well:

"Man who waits for roast duck to fly to mouth must wait very, very, long time."

Here is a more serious quote from Pablo Picasso:

> *"Action is the foundational key to all success."*

And finally, Brendan Burchard's insightful quote regarding the importance of action:

> *"All those who have won major life victories have found that all the resources needed to win are within, and that most knowledge needed to succeed is acquired after action."*

Let's look at the critical relationship between learning and action. Learning begins with identifying your knowledge gaps. How do you do this? One way is to examine your vision and goals and determine what new knowledge is needed to successfully achieve your goals.

You close your knowledge gap, for example, by reading books and articles; having a mentor or coach; interview experts; attend school; or take training courses.

Learning new methods, techniques, and tools does increase knowledge, and results in professional development. But, this is only the first, and easiest, step. That knowledge is not converted to an increase in intellectual capital or net worth unless the knowledge is deployed to do something that hasn't been done before. In other words, when you apply what is learned in the training, you achieve new goals, and grow. Learning brings development, but growth requires application of what is learned. This is *action.*

That brings us to the conclusion of the Consultant's Cycle of Success, the seven habits that appear to be closely related to success in consulting. Of the seven Cycle of Success factors, or habits, how many are currently part of your daily routine? Do you think that adding more of these habits would be beneficial?

Next up, we discuss the second key skill for consultants, which is business development.

III. Key Skill No. 2: Business Development

My experience with supervising staff in several consulting firms, from large to small, taught me that business development was the least understood skill for engineers, scientists, and planners. I believe this is because most technical consultants have been educated in technical schools, not business schools (although, even some business schools also do a poor job of teaching business development). Yet, business development is, perhaps, the most critical skill for a consulting firm. Similar to the importance of "taking action" in the previous chapter, without business, there is no firm.

Business development is a very broad term, meaning many things from marketing, to sales, to proposal preparation and to other topics that may be unfamiliar to engineers who lack business development training. Therefore, the subject of business development takes up about half of this book!

In this chapter, we introduce and provide an overview of a business development model we call The Compleat Business Development Model (spelling intended). We will explore each of the five components of the model. We dive into writing proposals and preparing presentations in Appendix A.

Maximum Influence

Figure 3-1

Our goal in applying the Compleat Business Development Model for Consultants is to achieve "maximum influence" over your prospects and clients. Figure 3-1 shows a B17 bomber. My father was a B17 pilot, and led 35 missions over Germany, often with up to 100 B17's in formation. This is what I call *maximum influence.*

The Compleat Business Development Model consists of a number of inter-related techniques that will increase the effectiveness of your marketing, and will utilize several persuasion techniques to increase the desire for your service offering. Maximum influence in marketing can only be achieved by employing *multiple* methods in your marketing, not just one or two.

Maximum influence, created by the Compleat Business Development Model, will be your strongest, most useful and potent marketing weapon when fully implemented.

SCT's Compleat Business Development Model

When we consider business development, three elements come to mind. First, we need a client or prospect. Second, we need a consultant who wants an opportunity to work for the client or prospect. And third, we need an offering, usually in the form of a proposal, linking the client and consultant.

Business development is a set of activities, such as lead generation and lead conversion. If we look at activities as processes in a system of processes, the process input comes from the consultant, the process is creating the offering, and the output goes to the client or prospect. There are great benefits to viewing business development as a system of processes.

Let's take a closer look at the process input, which is our consultant. First, we have the consulting firm and its benefits, such as work experience and staff expertise. We can position the firm favorably using these benefits.

Second, we have the staff member who will take the lead in pursuing an opportunity. This is typically a project manager. The project manager's ability to communicate effectively with the prospect is a critical element towards convincing the prospect to buy from you, as is the project manager's expertise and experience.

Third, we have our prospect. The prospect looks at the consultant through filters that include logic and emotion that make up the prospect's personality and how he or she communicates. We need to apply marketing psychology by employing psychological triggers and language that our prospect will respond favorably to and decide to buy from us.

Fourth, we have the offering. The proposal must address two important factors: the service offering customized for the prospect and the value as perceived by the prospect. In other words, we must create a winning value proposition.

These four elements, along with a systems approach (the fifth element), form the foundation of The Compleat Business Development Model depicted in Figure 3-2. We will take a deeper look at each of the five elements.

Systems in Business Development: The Real Advantage

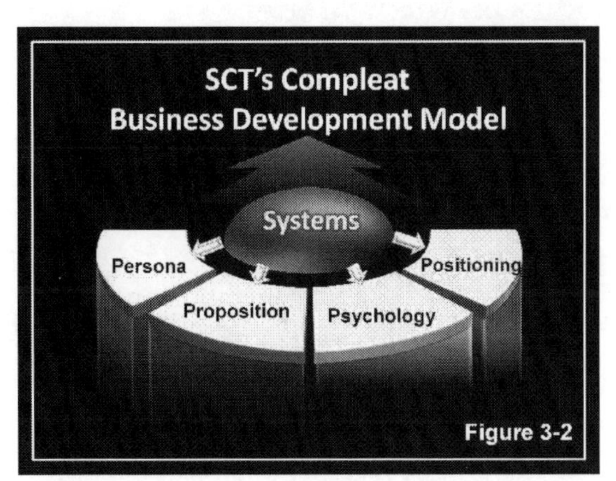

Figure 3-2

Looking at marketing as a set of processes rolled up into a system offers significant competitive advantages that are underappreciated. In this section, we begin with an introduction into systems thinking and the science of systematics, and we conclude the section with a suggested business development system.

Business Development as a System

Every consulting firm utilizes many systems. Each staff member uses many systems and processes. Doesn't it make sense to make these systems as efficient and effective as possible? Here is an insightful quote from my mentor, Dr. Deming:

> *"Your system is perfectly designed to give you the results you're getting".*

Read that quote again. Think about what Deming is saying. It is profound. Let's assume that your job is to generate leads for new business. Lead generation is a system with a number of separate processes, such as market research and lead capture. You probably have not looked at these activities as a system (a set of processes), and it is doubtful that you have optimized this system to give you the maximum output achievable with this system. As a result, you are getting the results your current system is designed to create. And you have no idea what the potential is for this system or whether you are achieving this potential.

Dr. Deming also stated that eighty percent of the effectiveness of your work is defined by the system you are working in. What he is saying here is that the system you are working within has a lot to do with your productivity. This is why you should consider a system approach to business development (and all other tasks!).

In summary, marketing through the use of systems and processes offers a significant opportunity for differentiation from your competition. In addition, if you are still not convinced, here are a few more advantages of using a systems approach in your marketing:

- When you look at the world through the lens of a system, you can easily identify opportunities for lower costs (more competitiveness);
- Improving a system will show you how to save time;
- A systems approach will provide significantly greater consistency, and this will show up in the form of higher proposal win rates and increased profits; and
- You will see fewer dropped leads.

So, why do so few consulting firms use systems in marketing? One reason is that is involves effort. To get started, you have to do a deep dive into your marketing processes and map them out one by one. Then you have to monitor their effectiveness. And finally, you have to optimize them. This takes some effort, and your clients are not paying for the time you need to undertake this effort. However, this is a long-term commitment. The time spent up front will pay off handsomely down the road!

Thinking About System Thinking

Let's look at how we think about and work with systems and processes. We will begin with processes. First, a couple definitions are in order. A process is a defined work activity, such as a "Go No-Go decision process".

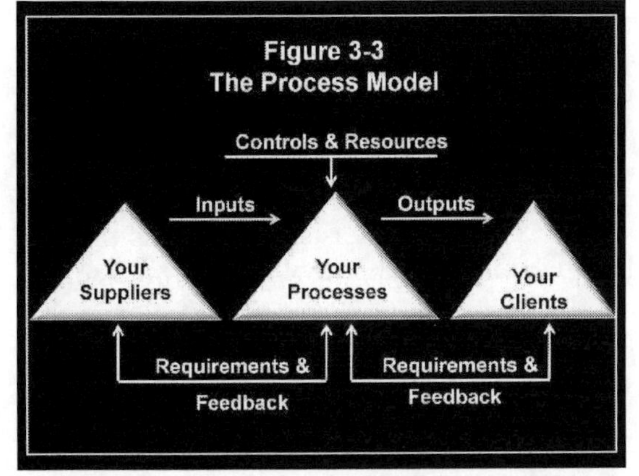

Every process has three components: an input, an output, and a value-added task. Figure 3-3 shows a "process model". Let's use the example of word processing to demonstrate how a process model works.

The input can be called "your suppliers". A process receives input from its suppliers. This can be materials, resources, or data. If you are a word processor, your suppliers are staff who provide typing assignments, for example, as well as the computer specialist who maintains your computer.

The work the word processors do for their clients (in this case, staff) represents the process. This process may include typing, copying, or other tasks as well.

This process has outputs which we call "your clients". In this instance, the word processor's clients are once again the staff, although indirectly the word processor's client is the firm's client.

Next, a process needs controls and resources. Controls are inputs that define, regulate, or influence the process. For the word processor, a control would be the Style Manual that governs how documents should look.

Resources include both human resources and physical resources that are associated with the process, such as the number of word processors needed for a shift, and the computers outfitted with the required software.

Next, a process must meet requirements, receive feedback, and provide feedback.

Requirements are characteristics specified by suppliers, process owners (e.g. word processing department manager), or clients to enable the process to produce the highest quality output. A requirement of the supplier would be due date. A requirement of the word processor back to the supplier (staff) would be "provide sufficient time to do the work". The scope of work is another requirement, as is the contract for the work assignment.

Feedback provides the process model with *stability*. In fact, we could call this process a *feedback control system.*

With feedback, you can answer the question "*is this process consistently delivering the output desired by the client?*" If the answer is yes, your process is stable and fulfilling its objective. If the answer is no, the process is unstable, and requires improvement.

Feedback is the most important part of the process model. Without feedback, a process will gradually become unstable. *Why do you think this happens?* There is no guidance system.

Every staff member should examine his or her process and determine if it can be improved. Here are a few questions that should be asked:

- Is the "voice of the client" being listened to?
- How can the process be improved?
- Can the process be simplified?
- Can scheduling of supplier materials delivery be improved to avoid delays?
- Can supplier and client relationships be improved?
- Can the client's needs be better understood?
- Can more value be added to the client?

Now let's go upstream and discuss systems. A system is composed of a set of processes. These processes are interconnected and interdependent. This means that actions by one process will affect all other processes in the system, and, perhaps, affect other connected systems at a higher level. The organizational structure, for example, depicts a collection of functions rather than a system of interacting, interdependent parts.

When thinking about systems, think in terms of nonlinearity or feedback loops. On the other hand, linear thinking implies that responsibility lies elsewhere. *If things go wrong, put the blame somewhere else.* When problems arise, they cross the boundaries of groups or divisions or teams in an organization.

For example, a problem in marketing should not be viewed as a marketing problem. The complex dynamics of a system must be examined.

Dr. Deming taught system thinking, and one of his main points was to pay attention to the definition of a system:

> *A system is a network of interdependent components that work together to accomplish the aim of the system. A system must have an aim. Without the aim, there is no system.*

A system without an aim is like being at a rifle range without targets. Without an aim, or goal, the processes will not work together well enough to achieve a higher result or reach the full capability of the system.

Let's look at a simple example of a system in a consulting office. Figure 3-4 shows a system comprising a business development team, staff working on service delivery, and a financial team monitoring billability. The sales target is specified by management.

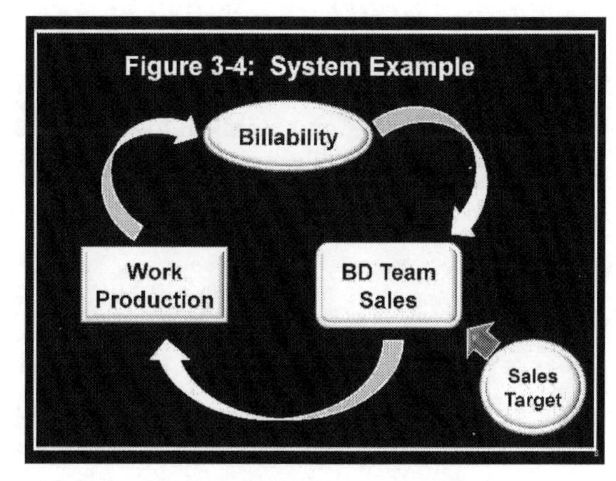

Management creates a sales target for the business developers. The business development team goes to work, making sales (hopefully!). As the work comes in, it is produced by project teams. *Now, suppose billability falls off? Why did that occur?*

Well, the problem could reside in the work production group, be the result of poor service delivery, or poor quality work. The cause could be due to ineffective sales efforts, insufficient marketing resources committed by the business development team, or perhaps management made the sales target too high, and the business development team became frustrated and deflated. Once again, system dynamics are at work!

Note that all of the processes are interrelated and even more importantly, *interdependent.* As a result, you just cannot look at one process in isolation. *Each process impacts other processes, and each system impacts systems above it or connected to it.*

Finally, the sales target is an important part of the system. Imagine what the results would be if there was no sales target. Do you think the results would be optimal for this case? No aim = no system.

Deming emphasized the importance of the system when evaluating the performance of an individual. The individual staff member continuously interacts with the system, and this must be considered, not only when assessing performance, but especially to optimize performance.

Systematics: Optimizing System Performance

Systematics is the science of creating and optimizing systems and processes. Let's assume that you have adopted a systems approach to marketing, and that you have created a business development system. What is the next step? Improve the system.

A system can be improved or optimized by carrying out one or more of the following tasks:

- Does the process have a clear purpose or aim? That aim should be consistent with the aim of higher order systems (e.g. office goals).
- Assign someone to each system to monitor the results and determine if the system is operating optimally.
- Gather data on the system or process output. Dr. Deming always said "Trust in God. Everyone else bring data." Without data, we cannot know whether a process is producing the results we need. We'll provide some examples shortly.
- Is the process output stable or unstable? This is a critical question. If the process is stable, it is producing at its design capability (which may or may not be optimal). Only common causes or structural causes of variation will exist in a stable process. On the other hand, if the process is unstable, some

special causes or what we call "tampering" will be responsible for creating erratic results with little consistency. We'll discuss process stability in more detail.

- Apply appropriate process improvement tools. Use process improvement tools to help a process perform optimally. Examples of process improvement tools will be presented shortly.

Gathering data for a system is essential if we want to understand and optimize a system. As an example, a few years ago, I was concerned because we were going over budget on a lot of projects. We collected data on this issue over several months. The results are shown in Figure 3-5.

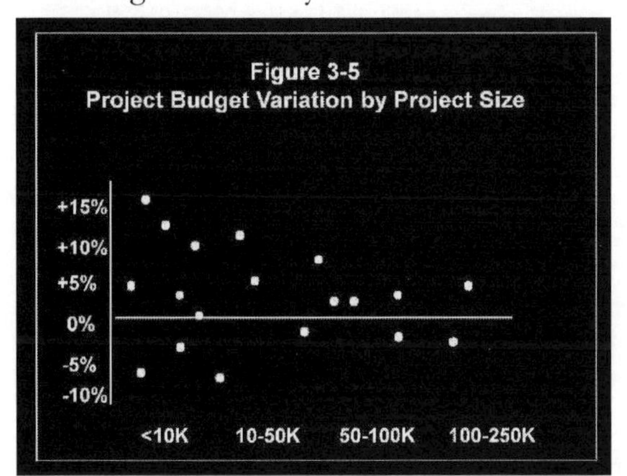

When we plotted the project variation versus the project budget size (stratifying the data), we realized that the greatest variation belonged to the smallest projects, an unexpected, but welcome result.

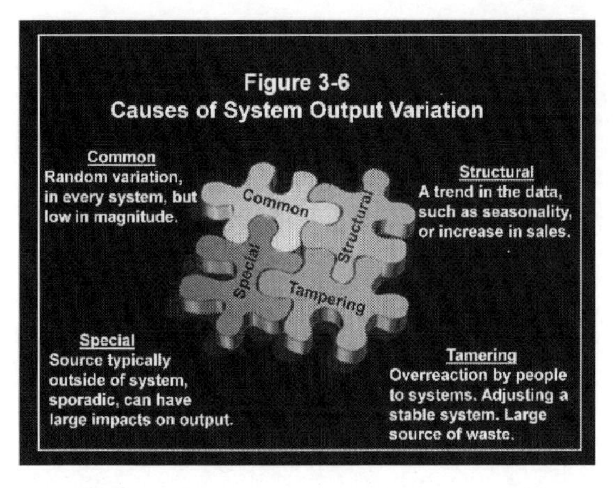

The importance of system stability was mentioned above. When we say stability, we are referring to the variation in output. There are a number of potential causes of variation in systems. Four of the more frequent causes are shown in Figure 3-6. Each of these is discussed below.

Common Causes of Variation: Common causes are created by randomness, typically at low magnitudes. Common variation is in every system and process. Examples might be changes in weather conditions on solar panel output, or an accountant taking a personal call while at work.

Special Causes of Variation: Special causes usually come from outside the system. They are sporadic, not continuous. They typically produce large impacts in system performance. Examples include a computer crash, a flood preventing employees from getting to the office, an unexpected cost, or a change in tax regulations.

Structural Causes of Variation: A structural variation typically shows up as a clear trend in the data, such as seasonality, a steady increase in sales, or the slow degradation of output from solar panels. This type of variation is typically "built-in" to the system.

Tampering: Tampering means someone overreacts to a situation, creating an unnecessary, large variation in output. The most common form of tampering is management asking staff for explanations of common cause variation. For example, thinking there is a trend (based on insufficient data) when there is no trend and demanding a response to the "trend". Another example: start and stop the marketing campaign based upon short term results.

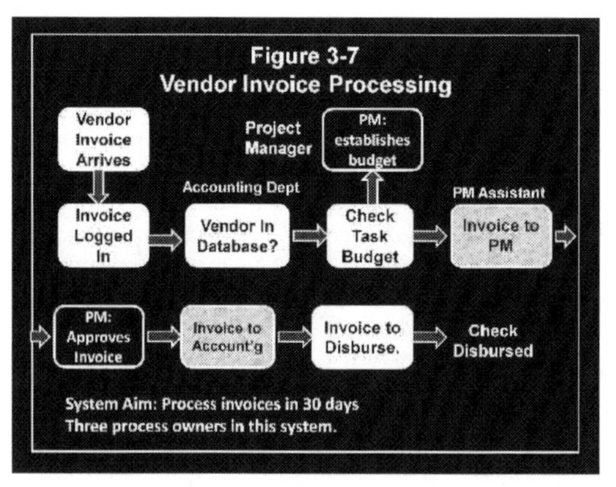

One of the best ways to determine if there is undesired variation in a system is to plot output data. An example is the control chart. Figure 3-7 shows a vendor invoice system.

The aim of the system was to process and pay an invoice within 30 days. At one time, we were getting complaints from vendors about late payments, which can trigger interest charges as well. The system has three process owners: the project manager, who must approve an invoice, the accounting department who processes the invoices and ultimately disburses the checks to the vendor, and the project manager's assistant, who helps the PM.

We compiled data from tracking 25 invoices through the system. A control chart of invoice processing days was created, and is shown in Figure 3-8.

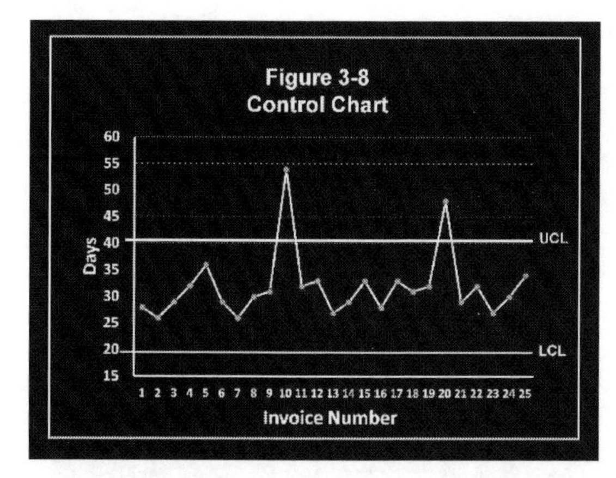

This control chart has an Upper Control Level (UCL) and a Lower Control Level (LCL), which show the limits of common cause variation. The data shows that common cause variation fell between 25 and 35 days.

The chart shows two "special causes" of variation, where the processing time reached 54 and 48 days. It was determined that the problem was a bottleneck at the project manager, because oftentimes they were out of the office, thus delaying invoice approvals. The solution was to give PM assistants more responsibility.

Since the range of common cause variation exceeded the 30-day aim of the system, one might ask what could be done to bring the common cause variation to within the 30-day requirement. This would require that the system be re-designed.

Lastly, there are a number of techniques that can be used to improve or optimize a system or process. A few of these include:

- **Simplification:** Simplify the system by eliminating processes.
- **Addition:** Add processes that will improve performance.
- **Sequencing:** Alter the sequencing or order of processes in the system.
- **Debottlenecking:** Look for areas where performance is hindered.
- **Training:** Provide additional training if personnel are not effective.
- **Charting:** Fishbone (Cause and Effect) Charts

The Business Development System

In this section, we'll examine the business development system. Let's begin with a quote from author Jull Konrath:

"In sales, systems outperform miracles."

The presence of a business development system is not unique. But, every firm has a different system. The key is to create a system that works for you, track its performance and seek to optimize the system over time.

One example of a business development system is shown in Figure 3-9. We will use this example to discuss a number of business development processes.

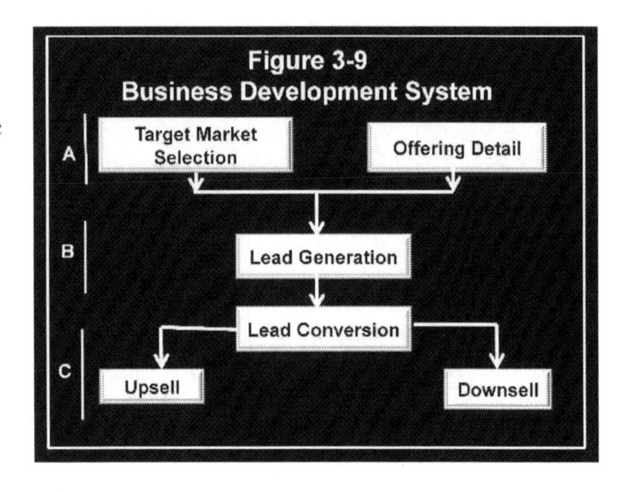

This business development system can be divided into three components:

- Initial Planning: The Client and The Offering,
- Lead Generation, and
- Lead Conversion.

Each of these components, and the processes in them, is discussed below.

Initial Planning: The Client and The Offering

The first component of this business development system deals with the prospective clients and the offering. We need an offering and we need a market. If you have an offering, you need a target market. If you offer many services, or you have the capability to customize services to a market, you need to conduct some market research to identify an attractive market.

If you have an offering, and you are looking for a market, I would ask the following questions. I need a "yes" for all five questions:

- **How well do you understand your service offering?** This sounds like a silly question, but it isn't. If you do not have a thorough understanding of your service offering, how can you sell it? For example, do you have a catchy name for the service? My expertise is site selection for the power industry. I called my service "Opti-Site" for optimal siting. The title gave me a segue into explaining the advantages of the service. Can you describe the service in terms of a clear framework? Steven Covey wrote about traits of successful people. He created a framework around seven habits. Can you relate exactly how your service offering will work for a client? Can you customize it so your client thinks it was developed just for him? Can you enumerate all the key benefits gained by using or applying your service? Are these benefits expressed in terms of your client's business drivers?

- **Is there a need for you service offering?** Think critically about this. Don't fall in love with your offering. The best offering in the world is useless without market demand. Do enough market research to convince yourself that the demand is there.
- **Will the marketplace recognize its value in your offering**? What does your value proposition look like? We discuss the value proposition in detail later in this chapter.
- **Will the market participants pay your fee?** How do you know? Will the prospect see the value?
- **Are they likely to use your service *now?*** The demand for your service offering must be soon, not a year into the future.

Use these questions to screen candidate target firms. If you are having trouble with these questions, do more market research. Here are several research methods that have proved very helpful for me:

- **Regulations:** Regulations create new business opportunities: Clean Air Act, Clean Water Act, Renewable Energy Portfolio Standards in various states requiring 25% to 100% renewable energy usage, and so on. Whatever you field is, look at the Federal and state regulations that influence this field and understand what regulatory changes are coming.
- **Legislated incentives:** Federal and state legislatures are always on the dole for their favorite projects in their jurisdictions. The

military spends enormous amounts of money on everything from new construction to renewable energy.

- **Market trends:** Your job is to learn about market trends in your field. What is the next great idea? How can you help your clients? By trends, we include social trends, demographic trends, economic trends, trends in retail, etc.
- **Clients in region:** Research potential clients in your area. Which ones are growing? What new firms are moving into your area?
- **Buyer needs survey:** Interview potential clients to determine what they want next, whether your service offering is a fit, and how you might need to redefine or adjust your offering to be in closer alignment with the prospects need. You can find potential buyers at business conferences and technical conferences.

When you bring your service offering to market, what is your strategy? For example, you can market to an industry sector, a services sector, or to a tightly targeted set of potential client firms.

Marketing via industry sector means we stay within one industry or sub-sector of that industry. For example, when I sell site selection services to the power industry, I can focus on one of several sub-sectors, such as investor-owned utilities, independent power producers, or municipal utilities. There are a number of advantages to staying within one market sub-sector:

- You can leverage both projects and people. Each industry firm will have similar needs for projects, use similar language, and so on;
- Referrals are easy to obtain if you do excellent work;
- You can become a dominant player within the sector;
- You can offer a basket of services that most firms will need;
- You will become very proficient at service delivery, thus reducing fees and increasing competitiveness; and
- You can easily network horizontally, across the market sector.

Marketing via the services sector means that you offer your service across market sectors or sub-sectors. For example, I could sell site selection services to the entire power industry, regardless of sub-sector. Or, I could sell site selection services across different sectors, such as mining, manufacturing, waste management, and so on. There are some advantages and disadvantages to this approach, such as:

- Marketing the services sector works well when your expertise is widely recognized, and there is strong demand for your service;
- As your recognition increases, you can work across industry market sectors, or other sectors, providing more target firms;
- On the negative side, there may be less client loyalty;
- Services tend to be more of the "one-off" kind, which may cap revenues.

Finally, we can adopt the tightly targeted sector approach. We target firms where we have a special relationship. We may have long-time friends who work at these firms. Or, the demand for your specialized service will be high. Or, these firms may be geographically attractive.

Client loyalty is a plus for this approach. However, leveraging your work may be more difficult. Also, revenues may be limited due to having a smaller targeted audience.

From my experience, the industry sector marketing approach offers the best advantages.

Lead Generation

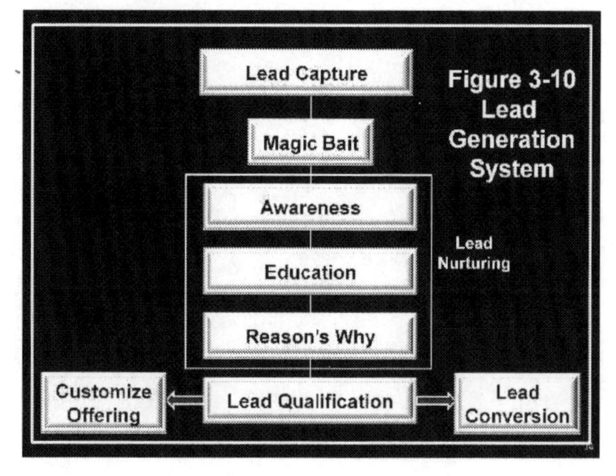

Figure 3-10 Lead Generation System

The second component of the business development system in Figure 3-9 is lead generation. One way to break down the Lead Generation System is shown in Figure 3-10 (recall that there is no "correct" or "unique" system).

Note that we keep breaking down the systems until we get

basic processes. The secret to a systems approach is to drill deeply into a system so that a full understanding of the system is attained.

Lead Capture Process

You have a target industry. How do you approach it? Select the right channels. The emphasis is on channels (plural). There are many advantages to using multiple marketing channels, however, most consultants only use one marketing channel (the one they are comfortable with). Using multiple channels will expose your service to more and *different* potential clients.

Activity in each channel varies with economic conditions and the type of service offering. Being in several channels minimizes revenue shortfalls. If one channel is soft, another may be strong. Always keep testing marketing strategies on each channel so you can optimize your return.

Examples of lead generation opportunities include:

- Network at conferences. This is a traditional, yet one of the very best, ways to network and get your word out. When you go, be sure to get a copy of the attendance list. This is a *great* lead generation source!
- Advertise offline. Advertising is still a common method of marketing. You can advertise in trade journals, for example, or professional association newsletters.
- Advertise online, LinkedIn and Facebook are excellent places for online advertising, even for technical firms.
- Participate in professional societies or associations, especially ones where your clients are members.
- Attend city council meetings, hearings, and civic organization meetings.
- Secure conference attendance lists. Get them for the conferences you attend. Scour them for leads. Try to obtain attendance lists for other good conferences. If they are not available for a reasonable cost, go to the conference's website, which usually lists registered attendees before the conference begins.

- Attend trade shows relevant to your client base. Staff a booth for more impact. Even better, have a drawing for a good prize. Collect business cards and ask that a short form be filled out with a few qualifying questions to get a drawing ticket.
- Send direct mail/email campaigns. This is the classic method. Direct mail is still active, based on the number of mail pieces I get every day. You can purchase very focused marketing lists. Just Google "sales leads" or "sales lists".
- Form alliances with non-competing firms. Alliances, or affiliate partnerships in Internet parlance, are a great way to build your business. Find complementary services and piggyback on their marketing list. Interior decorators or furniture stores can set up deals with real estate agents so they become aware of recent house sales. The interior decorator shares their fee with the real estate agent. There are many similar situations where two firms can share resources.
- Publish papers and speak at conferences or trade association meetings. Publishing and speaking are excellent ways to get your name out, and it positions you as the expert. When you decide to attend a conference, pay attention to the attendee list, which is typically published in advance to attract more people. Use that list as a contact list while at the conference. Use the final attendee list as a contact list after the conference. Attend all lunches; they are great places to meet people. Talk to key speakers right after their talks. Get as many business cards as possible. If you cannot attend a conference, try to purchase the conference materials, which should include the attendance list.
- Publish a newsletter. Today, it is very easy to publish an online newsletter that can be emailed to a list of potential and existing clients. A simple newsletter can be the source of significant business opportunities.
- Engage past clients. Past clients can be a great source of leads. This works even better if you publish a simple newsletter that contains useful information about the client's business. By keeping in touch, the likelihood that the past client will provide an opportunity is much higher.

- Hold technical transfer webinars. Webinars are excellent ways to attract the attention of potential clients and get them to participate. Select a topic that is of interest to a wide variety of prospects. Or, engage an expert or government agency staff member to talk about an interesting topic.
- Attend bidder's meetings. Bidders meetings are terrific venues to meet clients or potential alliance partners. Attend bidder's meetings for the purpose of meeting people, not just to learn more about a proposal opportunity.
- Establish a presence on LinkedIn. LinkedIn is a great way to connect with people that should be on your contact list. All of your employees should have a profile on LinkedIn. The company search and name search tools are excellent. Try the Sales Navigator and use LinkedIn Groups
- Get to know your vendors. Many of the vendors you use also work for your potential clients. They can help provide contacts.

In summary, brainstorm possible channels to collect contacts for potential clients, and pursue multiple channels.

The Magic Bait Process

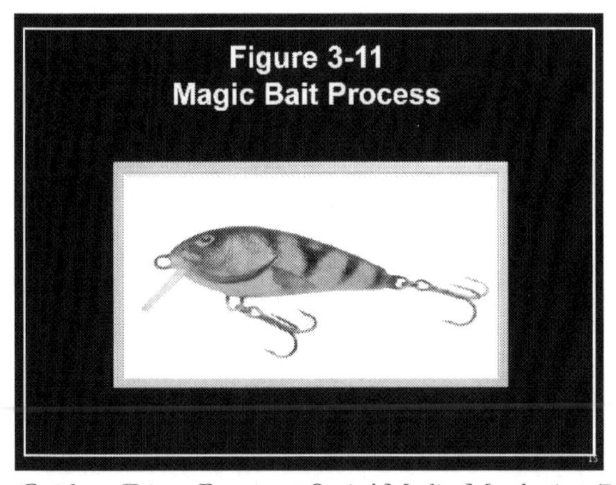
Figure 3-11
Magic Bait Process

One kind of magic bait is shown in Figure 3-11. Magic bait is used by many marketers, and has been shown to be a valuable technique. Author and marketing expert Dan Kennedy emphasizes the importance of this technique, which he calls a "lead magnet", in his book *No B.S. Guide to Direct Response Social Media Marketing* (Kennedy & Walsh-Phillips, 2015). Consultants should use this technique more often.

Magic bait means that we "lure" a prospect to inquire about your service offering. Or, we "set the hook" with a potential lead. Magic bait is something of value that you give to a prospect to get attention or position yourself as an expert.

Magic bait takes advantage of a psychological trigger called The Law of Reciprocity. Robert Cialdini wrote about using triggers to influence buyers in his landmark book *Influence: The Psychology of Persuasion* (Cialdini, 2006). Cialdini says The Law of Reciprocity obligates a person to repay what another person has provided. So, by sharing something of value with your prospect, he/she feels obligated to return the favor. We talk more about psychological triggers later in this chapter.

Examples of magic bait include:

- A free report about something of interest to the prospect
- A newsletter issued periodically
- A recent article of interest
- A recent and relevant case history
- A research report prepared by the government or other entity
- A press release
- A link to a third-party report
- A live or recorded webinar on a technical topic of interest
- Breakfast, lunch, or dinner
- A golf outing or fishing trip
- Attendance at a sporting event
- A conference paper you wrote
- Reference to a blog post
- An inexpensive, useful gift

Take advantage of the Law of Reciprocity. It is a powerful source of influence.

The Nurturing Process

Figure 3-12
The Nurturing Process

The next step in the example Lead Generation System is the Nurturing Process (Figure 3-12). Early-stage leads are in a very fragile state. Both sides are carefully evaluating each other. We need to handle these leads with care!

Nurturing is defined as "building relationships one step at a time". The purpose of nurturing is to maintain a conversation with the lead until the lead is "sale-ready". Research indicates that as many as 90% of leads just entering your lead capture process are not sale-ready. However, nearly two-thirds will eventually be sale-ready if a nurturing process is used.

To help you better understand the nurturing process, I've broken down the process into four parts:

- **Awareness.** Once a lead enters your sales funnel, or database, the first step is awareness, where you welcome the lead, and inform the lead about your firm. This step is part of the positioning process, which is discussed later in this chapter.
- **Education.** Here, you begin to educate the lead about your service offering, and provide industry-oriented, relevant content. This step also is part of the positioning process discussed later in this chapter.
- **Reasons Why.** In the third step, you describe why the prospect should engage you. Now you talk to your prospect about the benefits associated with your offering. The benefits are identified in the value proposition process described later in this chapter.
- **Why Not.** The fourth and final step in the nurturing process is where the objections are identified, and responses are crafted to overcome them.

It's important that you follow the four-step process described here, one step at a time. Don't combine them into one conversation with the prospect.

The Lead Qualification Process

The final process in the Lead Generation System is the Lead Qualification Process. You've brought a lead through the nurturing stage. Now is the time to qualify them for the sales process. Or, not.

There are two parts to this process: first, gain an understanding of the buying criteria, and second, now that you have some understanding of the opportunity coming up, begin to customize your service offering to the specific opportunity. Note that these two steps must be completed before a Request for Proposal is issued. This is the last step in the Lead Generation System.

The key buying criteria are:

- What is the budget for the opportunity?
- Who has the authority to decide on a consultant for this opportunity?
- When is the Request for Proposal, or equivalent, going to be released?
- What are the specific explicit (overt) needs or concerns?
- What are the implicit (covert) needs or concerns?

Tracking the lead qualification process can be useful, since it focuses attention on the process. I employed the following tracking system in the past. One "qual" was awarded when each step was successfully completed.

- One-way communication established (either an inquiry from the prospect or you reached out to the prospect). The first qual is earned.
- Two-way communication is established (this is the Awareness Stage). The second qual is earned.
- Acquaint the prospect with your firm's qualifications in detail (this is the Education stage). The third qual is earned.

- When you provide the Reasons Why your service is best for the prospect, you earn a fourth qual.
- When you learn about objections (the Why Not stage), you earn the fifth qual.
- Finally, when you discover the buying criteria, you earn the sixth and last qual.

I used to run contests among the seller-doers in the office. Cash bonuses were awarded for earning a certain number of quals in an office marketing campaign. More importantly, however, this allows you to accurately track your success and look for ways to improve your results. Bottom line, you do need some sort of metric or monitoring device to follow leads so that you recognize when to ratchet up the selling effort.

Lead Conversion

The last component of the example business development system we are discussing is Lead Conversion, which includes writing proposals and giving presentations to win assignments. For the purpose of this example, we will assume that a proposal must be prepared. We define a Proposal Preparation Framework comprising the following five steps:

Step 1: Planning. The planning activity includes a number of tasks, including gathering proposal intelligence from your client and others. Also, you want to begin thinking about sub-consultants or contractors you may need to round out your team.

Step 2: Strategy Development. This activity includes putting together a winning project team, selecting the Project Manager, describing the service and how it will be applied, and listing *key differentiators* (to thwart your competitors) and *hot buttons* (to attract the client).

Step 3: Design. In the design activity, the proposal document is outlined to ensure that the needed support is available for word processing, graphical design, etc.

Step 4: Production. Next, the proposal is produced. Use a detailed checklist to ensure that all required tasks are carried out. Make sure the document is logically formatted so that the reader can follow your story.

Have an editor review the document for grammar and readability. Conduct a page turn after the document is compiled (before copying). Also, have someone not involved in its preparation review the document.

Step 5: Post-submittal. After the submittal, you must stay in contact with the client and seek any feedback. If a problem is identified, suggest ways to overcome the concern.

Step 6: Proposal Debrief. After the proposal effort has been won or lost, conduct a debrief. First, look internally and criticize each step in the Proposal Preparation Framework. Second, interview the prospect to determine what they liked about the proposal and what concerned they had. Create a written record of each debrief. Use this data to continuously improve the proposal preparation process.

In Appendix A, we will hone in on the preparation of proposals and presentations.

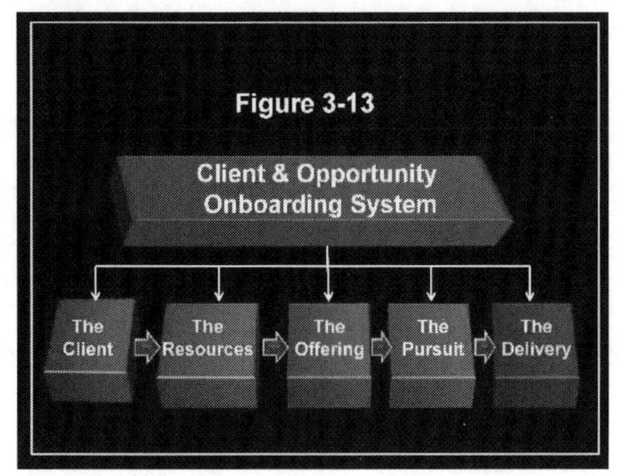

Once again, we used a simple example of a business development system to show how you use systems and processes in marketing. In closing, Figure 3-13 shows an overview of a more realistic business development system we've used.

There are many processes conducted within the five sub-systems. For example, The Pursuit includes Positioning, Lead Generation, Lead Nurturing, Lead Qualification, and Lead Conversion. Some of these processes can be further divided. When designing a business development system, processes must be created within each sub-system. A complete business development system can have 20-30 processes.

With a business development system in place, the effectiveness of your marketing and sales efforts will be markedly improved. Tracking progress becomes easier. What works and what doesn't work stands out.

Persona: Harnessing Our Differences

Next, we begin to discuss the four components of the Compleat Business Development Model: Persona, Value Proposition, Marketing Psychology, and Positioning. This section addresses Persona, or how we identify our personal differences and use them to create competitive advantage.

Sally Hogshead, author of *How the World Sees You* (Hogshead, 2014), says each of us has a responsibility to provide our highest distinct value to our clients. You provide your highest distinct value when you describe how you are different and what you do best. Let's repeat that statement:

"You provide your highest distinct value when you describe how you are different and what you do best."

Harness Your Personal Difference Advantages

What is a Personal Difference Advantage? Today, being better isn't enough. You must have a Difference Advantage. When you are different, you stand out. When you are different, you get attention. When you are different, you influence others. When you are different, you win. The key is to recognize and utilize your *differences*, not just your strengths.

There are three ways that you can be different: your technical specialty, your persona, and how you communicate with others.

Your Technical Difference Advantage

Your Technical Difference Advantage is what you do best. Almost every consultant has at least one technical specialty. Here are five ways to enhance your value using your Technical Difference Advantage:

- First, tell people about your expertise. Your expertise is part of your brand. For example, I would say my expertise is consulting to the power industry.
- Second, tell people exactly what you do with your expertise. How do you help your clients? I would say I am an expert in site selection in the power industry.

- Third, what is your specialized offering(s)? I would say Opti-Site, my site selection process.
- Fourth, share examples of your innovative solutions. I would show how Opti-Site uses decision analysis, which enables us to optimize the site selection process.
- Fifth, be ready to share your targeted experience. I would describe one or two challenging, but successful site selection projects.

In summary, look for ways to use your Technical Difference Advantage to show you are different than the competition.

Your Persona Advantage

By "persona" we mean who you are, or what characteristics you possess that clients will be attracted to. Here are five such characteristics:

- You have your credibility, which is based upon your education, your licenses and certifications, and your awards. Emphasize your credibility.
- You have authority, which is primarily derived from your position (project manager, vice president, etc). Authority is very influential.
- You have recognition, or what others know about you. Emphasize what you are known for in your business.
- You are a celebrity. This is an extension of recognition, and means how well you're known. If, for example, you have a national reputation for being a recognized expert, that will help you influence your prospects and clients more than anything else. Think about how celebrities influence people.
- You have longevity as a consultant. Longevity is related to authority. People with a long work history are believed to be authorities.

Take advantage of your Persona Specialties. This will differentiate you from your competitors and demonstrate your uniqueness to your prospects and clients.

Your Communication Advantage

The third difference advantage is communication or, more precisely, how we communicate with others. How we communicate, or how our personality is observed by others, can be an important determinant of success, particularly in consulting., In *Talent Delusion: Why Data, Not Intuition, is the Key to Unlocking Human Potential* Tomas Chamorro-Premuzic says "career advancement is a function of how people see you", (Chamorro-Premuzic, 2017).

The Communication Problem

The Carnegie Institute of Technology performed a study of successful business people. They found that 85% of their financial success was due to their skills in communication and their personality. Only 15% of their success was due to their technical expertise. Therefore, we can conclude that how you communicate is an important factor in your success.

The uncomfortable truth is that most of us cannot see how we are perceived by others, such as prospects, clients, or supervisors. Without the ability to consistently and accurately communicate with others, none of us can truly succeed. We must take an active role in shaping how we can communicate effectively.

Here is a great quote by Heidi Grant Halvorson:

> *"You cannot simply sit back and expect other people to do a better job of judging you accurately. You are going to need to get actively involved."*

We must learn how to get the attention of the other person (e.g. the prospect). We must provide additional information about who we are. Otherwise, the other person (e.g. our prospect or client) will fill in the blanks themselves, creating a profile for you that likely will not be accurate.

This is where the science of Fascination comes in.

Introduction to the Science of Fascination

Most people, especially technical people, don't have a clear understanding of how to best communicate with influence, and therefore, don't understand how we add their highest distinct value. In other words, it isn't easy to recognize how we are seen, or understood, by our prospects, clients, team members, and supervisors.

We communicate in two ways. The first one is a commodity. The second one provides our highest distinct value. This is our competitive advantage. We stand out; we get attention. We influence others, including our prospects and clients.

To grow your business, you don't have to change who you are. In fact, quite the opposite. You must become more of what you already are.
Sally Hogshead

Sally Hogshead founded How to Fascinate, and is the author of the best-selling book on fascination, *How the World Sees You* (Hogshead, 2014) I am a Certified Advisor for Fascination, and received training from Sally Hogshead.

We must learn how to communicate with an intense focus, called "Fascination". When we fascinate, we are communicating at our best. When we fascinate, our communication is in a state of "flow". Our communication becomes a core advantage. When we fascinate others, we get their attention and we become much more engaging and influential.

When we fascinate someone, they're intensely interested to learn more about us, our firm, and our offering. Fascination happens because we are performing at our best, with an emphasis on our natural communication strengths.

Our Communication Advantage

To be a good marketer or salesperson, we need to be persuasive and influential. This is what Fascination is all about. When we fascinate others, we create a communication advantage, or communication difference advantage. To become more persuasive, we don't need to artificially change who we are. But we do need to identify and activate our communication difference advantage. This is linked to how we communicate at our best, and how we provide our highest distinct value.

For a moment, let's go back to 1985, when Edward de Bono introduced his Six Thinking Hats creativity aid. De Bono defined six ways that people approach situations or problems:

- A person who loves to rely on facts and is a logical thinker wears a white hat.
- A person who loves to innovate by generating new concepts and ideas wears a green hat.
- A person who uses intuition to make decisions wears a red hat.
- A person who process-oriented and is very organized wears a blue hat.
- A person who loves to make everyone happy wears a yellow hat.
- A person who is cautious, and likes to point out what will not work wears a black hat.

This model demonstrates that people communicate in different ways, from different points of view. People see the world differently. For example, engineers tend to wear white hats because they are comfortable with data. Managers typically wear blue hats because they are comfortable with processes and project management. With this background, we will introduce Fascination.

In the science of Fascination, Sally Hogshead discovered that there are seven communication advantages, or languages people use to add value. But only one or two of these languages will be most effective for a specific individual. When these advantages are used, the individual becomes more authentic and influential.

You could use any of the seven Advantages, theoretically, but there's going to be one that's most effective for you. This Advantage is the way in which you most effortlessly and effectively communicate."
Sally Hogshead

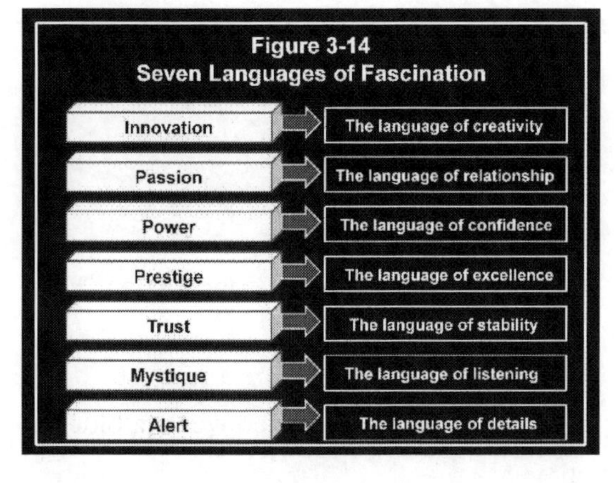

The seven languages of Fascination are shown in Figure 3-14.

Once we know a person's communication advantages (fascination advantages), we know a lot about how they communicate and "what makes them tick". Let's assume that you have determined that a certain prospect's primary communication advantage is Power. That tells us a lot about the prospect:

- **Influential:** Often looked at for guidance, answers, and assistance. An authority. A leader.
- **Opinionated:** Has strong beliefs. Known for their candor. Effective salespeople.
- **Decisive:** Makes decisions easily. Sets the path for others. Likely risk prone. Poor planner.
- **Confident:** Confidently confronts problems. Communicates with intensity.
- **Goal Oriented:** Strong drive to succeed. Sets high standards of achievement.

You can start seeing the power of Fascination. But we can take this even further. You have an upcoming meeting with the prospect, or you must prepare a proposal or presentation for the prospect. How do you approach it? Based upon the Power communication advantage, we can offer the following advice:

- Share insights and ideas with your client
- Stay focused and on agenda
- Use a clearly defined game plan
- Sharing pleasantries doesn't help
- Let the client talk about objectives
- Power clients value their time
- Ask about goals and show how you can meet them

- Offer options
- Offer bullish forecasts and comparisons
- Don't use urgency trigger
- Don't dominate the conversation
- Don't come across as uncertain
- Don't be aggressively proactive.

How do you think that this advice would impact the quality and focus of your proposal or presentation? Can you begin to see the value of a communication advantage? In this example, we used a prospect. However, knowing your communication advantage is equally helpful, and will enable you to be more influential and persuasive. We will demonstrate that in the next section.

Engineers score high on Mystique. Most engineers prefer to think through problems and processes independently, rather than participating in rowdy group brainstorming."
Sally Hogshead

Your Communication Archetype

Next, let's see how *your* communication advantage can help you in marketing and selling. The first step is to take a Communication Assessment and receive a Communication Advantage Profile. This is a short questionnaire processed by the How to Fascinate organization. The Communication Assessment is unlike Myers-Briggs or Strength Finder, or similar personality tests. This isn't about personality. It is about how you communicate, and how others (e.g. clients) see your communication. Instructions about how to get a Communication Assessment (also called a Fascination Assessment) are included in the Afterword.

The Communication Advantage Profile (also called a Fascination Advantage Profile) will tell provide your primary and secondary communication advantages, or your "communication archetype". There are 49 different archetypes, corresponding to the seven fascination languages.

We'll use my Communication Advantage Profile as an example. The Profile says that I am a Guardian (one of the 49 communication archetypes). The primary and secondary advantages of a Guardian are power and trust. These are my most effective modes of communication. Others see me as:

- A sure-footed leader
- Someone who sets ambitious goals for growth
- Someone who expresses opinions with a quiet unwavering conviction
- A strong reputation

My profile also provides a number of adjectives that reflect who I am, and should be used in marketing copy. Examples from the profile are:

- Prominent (solid reputation)
- Genuine (authentic, willingness to help)
- Sure-footed (confidence, focused on continuous improvement)
- Constant (stable, reliable)
- Resilient (unfazed by tough challenges, slow to tire)

My Communication Assessment Profile tells me how to describe myself when I am at my best, providing my highest distinct value to clients. To test the accuracy of the profile, I showed my results to co-workers and friends. Everyone agreed that this is how they see me!

The adjectives, and other information provided in the 15-page Communication Assessment Profile gave me a number of advantages I should remember to use in my marketing, such as:

- Demonstrate confidence and optimism that solutions will be found, and they will work
- Emphasize prominence by writing papers and speaking at conferences
- Remind others about the power of continuous improvement
- Talk about the relationship between goals and growth
- Use my expertise in decision analysis to help clients make effective decisions
- Emphasize my logical, step-by-step approach to problem solving

The Communication Profile provides your primary and secondary communication advantages, along with a number of adjectives that describe who you are at your best. When you communicate with your primary advantage, you become more engaged, more productive, more influential, and more persuasive. Your brain is less stressed or anxious. Words flow more easily. Communicating feels more natural to you.

You will not have to spend more money on marketing. You will not have to hire different people. You will need to learn how to leverage the advantages of your staff.
Sally Hogshead

And, don't forget the value of communication advantages when facing a client or prospect for which you have a relationship, and can guesstimate their primary communication advantage. As we showed earlier for the Power client, if we know a client's primary advantage, this gives us great insight into how to sell to them more effectively.

The communication archetype is quite useful when forming teams, such as project teams. Teams work best when there is diversity of personalities, and the personalities are known. In other words, teams with different communication styles will work better together, and teams with similar communication styles will tend to be dysfunctional. For example, if most of the team members have the Alert primary advantage, the team likely will become anal retentive and will micromanage. If all of the members have Power as their primary advantage, the egos will stall progress. On the other hand, members with a Passion advantage will enhance team cohesion. Members with the Mystique advantage will keep the team focused.

Based on the above, it is helpful to have all the staff get a Communication Assessment Profile. First, they will enjoy the experience of learning more about themselves. Second, they will learn how to communicate more authentically and be more persuasive. Third, the communication advantages provide management with a great tool, for teamwork and other valuable applications that are discussed shortly.

Difference Marketing

Now let's put everything together and apply difference marketing. Ask yourself the question posed in Figure 3-15: Who are you? Why are you different and what do you do best?

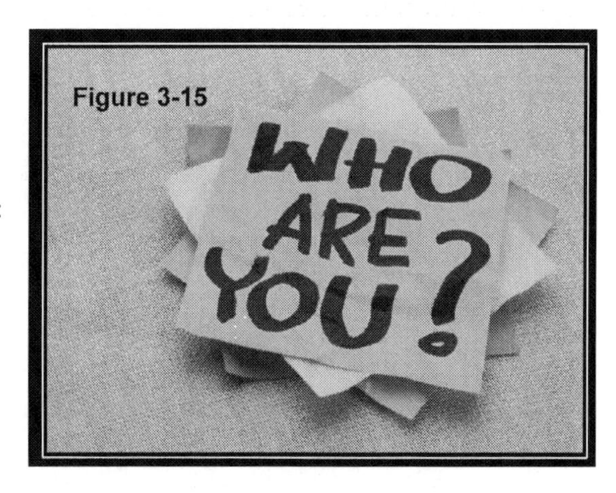

Figure 3-15

You want to use all of your difference advantages:

- **Technical Difference Advantage.** Clearly explain your expertise and how you use your expertise to create beneficial solutions for your clients. Describe case histories where you have successfully applied your expertise and offerings.
- **Persona Difference Advantage.** As we explained earlier, mention your credibility (education, licenses, awards), authority (position, committee assignments), recognition (what others know about you), longevity (years in business), and celebrity (how well you are known).
- **Communication Difference Advantage.** Use the communication archetypes as individuals, and apply communication advantages to your prospects and clients, as explained earlier. This is a powerful tool that will help you write winning proposals and give winning presentations.

When you are writing a proposal, preparing a presentation, writing a sales letter, or meeting with a client, don't forget to include your Technical, Persona, and Communication Difference Advantages. Remember, being different is more advantageous than being better.

Management Applications for Communication Archetypes

Communication archetypes are useful management tools. Here are three ways to use communication archetypes:

- **Project team assessments.** We discussed teams above. Once again, knowing the communication archetypes of team members helps create superior performing teams, and helps explain team dysfunction.
- **Performance evaluations.** Communication archetypes are quite helpful for annual performance appraisals. First, the communication archetypes help managers understand staff communication styles. Second, managers will know the natural strengths of their employees. Third, managers can avoid giving staff goals that conflict with their communication archetypes. Fourth, and most importantly, managers can give staff goals that match staff's natural strengths. Fifth, managers should assign roles and responsibilities that are consistent with their communication archetypes. When this is done, a win-win situation is created between staff and management, and the employee will be fully engaged because their responsibility is aligned with their strengths.
- **Hiring strategies.** Perhaps the most valuable application of communication archetypes is related to hiring strategies. For example, what advantages does your office lack? What types of people are most likely to deliver the results you need? A strong suggestion is to ask a candidate to take the Communication Assessment questionnaire and get the Communication Profile. You will learn so much about the candidate. However, that's not all. Suppose you want to hire a project manager. Have your two or three best project managers take the Communication Assessment and get a profile. Now compare the archetypes of your best project managers with the candidate's archetype. If you have a match, your new hire will be a very successful one!

Successful Consultant Training (SCT) offers a live online webinar on the topic of The Difference Advantage System. Details can be found in the Afterword.

The Underappreciated Value Proposition

Next, we discuss the much-maligned value proposition. Yes, we need to provide value to the prospect or client. However, in my experience, the value proposition is underappreciated and misunderstood. First off, a value proposition is *not* the same as a unique selling proposition. A unique selling proposition comes from the perspective of the consultant (seller), which is flawed. The prospect does not care what you are selling. They care about what benefits they will get and whether these benefits are consistent with their needs and wants. The value proposition, on the other hand, is a very powerful concept, and a powerful tool in marketing.

The Value Proposition Defined

Let's begin with the definition of a value proposition.

"The value proposition is a distinct benefit, superior value, or powerful result you offer that makes your services profoundly more attractive than your competition in terms of fulfilling the needs of your client".

Jay Abraham

The key words in this quote are "distinct benefit", "superior value", "services", "competition", and "client". The value proposition addresses a number of issues, but especially the benefits to the client. Here is another excellent quote:

"A value proposition is a clear statement of the specific, tangible results a client gets from using your service. Its outcome should be focused and stresses the business value of your offering."

Jill Konrath

Konrath emphasizes the importance of addressing the business value of your service offering

So, a value proposition is a clear, specific and differentiating statement of the actual results your client perceives they will get when they use your service offering. The key word here is "perceives". The value proposition is all about the client's perspective.

Before moving on, let's discuss the definition of the word "value". Value can be defined as benefits minus cost, as shown in Figure 3-16. While many definitions of value fail to include cost, we must address it. If you provide a benefit at a very high cost, the client will not see the value.

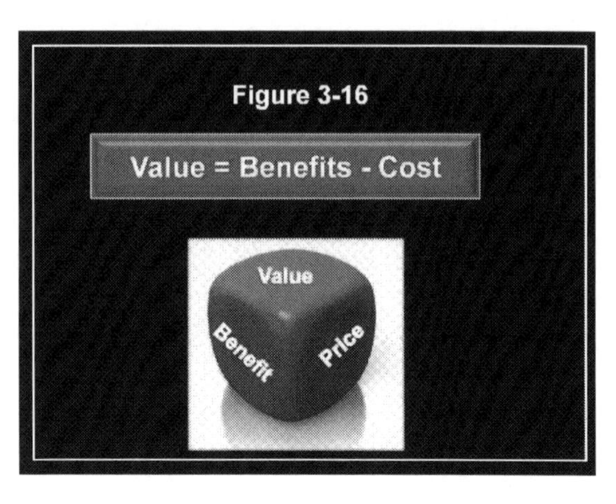

Benefits are the specific outcomes generated by your service offering. Benefits are what the client gets. Cost is what the client will pay for, such as your fee, time (schedule), and perceived risk.

But this is only part of the definition for value. When we think about value from the perspective of the prospect or client, we see that value has three dimensions:

- **Economic.** Economic measures are what we mostly see, including cost, reduce cost, increase revenues, increase profit, grow sales, or create a new profit center.
- **Emotional.** Emotional values include reducing anxiety, giving recognition, achievement, and employee engagement. For example, reducing anxiety, also called managing risk, can be a very important value in proposal evaluation.
- **Functional/Technical.** Functional or technical values include reduce risk, save time, simplify processes, process improvement, optimization, compliance, the use of best practices, and reduce effort.

It is important to consider the three interpretations of value when crafting a value proposition.

Crafting Your Value Proposition

So, how do we craft a value proposition? Begin by answering the following questions:

First, do we have an understanding of the client's problem? Or, can we see an opportunity to present to the client?

Second, what is the solution to the problem? The solution must be customized to fit the client's problem! What makes the solution different from that of your competition?

Third, what is the value gained by the client and the benefit to the client? We talked about the ways value can be interpreted in the previous section. Which of these is appropriate? This is essentially the benefit the client will see.

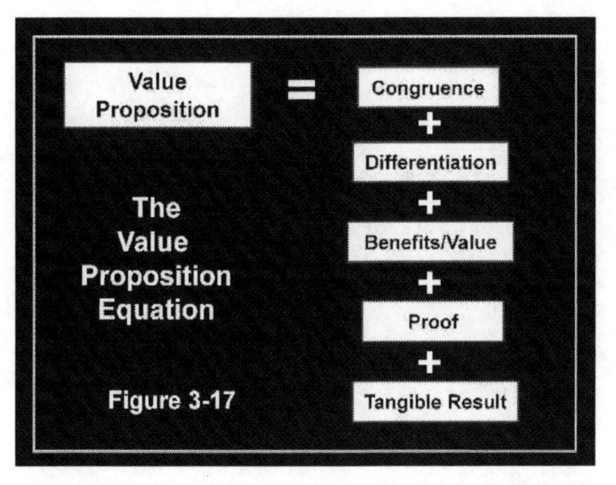

Figure 3-17

Fourth, what is the directional metric? A directional metric has two dimensions: a direction and a metric. For example, "our service offering will increase your revenues by 20% per year".

With these questions answered, apply the Value Proposition Equation, as shown in Figure 3-17.

Let's define the terms in this Value Proposition Equation.

- **Congruence** means that your service offerings are compatible with your client's needs, wants, or objectives. Focus on the outcomes, or results, that will provide the greatest benefits, or value, to the client. Focus on both overt (personal) and covert (firm related), issues.

- **Differentiation** means that your service is not just *different*, but *better than the competition.* Examples of differentiators are technical solution; contractual advantage; management or project management (e.g. budget control); better service; better facilities, equipment; more recognized experts.
- **Value** means you are increasing revenue or profit, reducing cost or risk, increasing market share, making operations more efficient, making management more effective, or reducing anxiety.
- **Proof** means support your claim of value with specific examples, such as a case history or testimonial.
- **Tangible Results** means you are using a specific metric related to your client's key business drivers. Examples of tangible results include provides a competitive advantage, increase a revenue stream or create a new one, reduce costs, reduce business risk, improve regulatory compliance or improve operational efficiency or effectiveness.

Clearly, there are instances where a specific metric cannot be identified. Here is an example where we needed a value proposition for construction of a walking path requested by a town council:

- <u>Level 1:</u> You describe your offering and along with it, you describe a few features of the offering. From our example, you might say your offering "a five-foot-wide walking path covered with staymat, as prescribed in the RFP." This is the least attractive alternative to a metric.
- <u>Level 2:</u> Now you present a key benefit to the stakeholders. In our walking path example, we might say "The public will love a walking path". This is a benefit, but weak because it does not strongly appeal to the decision-makers.
- <u>Level 3:</u> Next, you include a benefit to the firm or agency you are proposing to. For example, "The walking path will help the Town Council get re-elected." This is better.
- <u>Level 4:</u> Now you include a benefit for your contact at the firm or agency. For example, "The bill's sponsor will get his bill passed, and will gain respect and leverage on the council." This benefit is addressed to the key decision-maker.

- **Level 5:** Finally, you describe a covert benefit to your contact. What does your prospect want to accomplish personally regarding their need for services? Everyone likes to express their character, and everyone wants to gain recognition for their accomplishments. Does he want to achieve something? Learn something? Be recognized as an expert? Document their success? Increase their prestige? Can the solution or deliverable become a symbol for success for the prospect? Covert benefits are typically strong motivators.

In summary, think hard about what will motivate your prospect to purchase your service offering. The closer you get to covert reasons the better because you appeal to the emotion of the buyer. This is a solid alternative if you cannot find a suitable metric.

Example of a Value Proposition

Opti-Site International (OSI) was engaged by a developer of utility-scale photovoltaic solar energy facilities to locate a site for a 500-acre 80 MW solar project in the southeast. OSI created a value proposition for the purpose of submitting a proposal to the developer.

- **Congruence.** This means meeting your client's objectives. In this example, the client's objectives were a) they wanted a single-axis ground mount tracking system, b) required infrastructure (primarily transmission lines) must be available adjacent to or on the site, c) the landowners must be supportive of the project and willing to lease or sell their land to the project.
- **Benefits/Value.** The key benefits are a) Opti-Site, the siting process OSI uses, will identify the best, or optimal site in the study area, b)the site will be permittable, and should not experience excessive permitting delays, c) the siting process is rational and transparent, and d) the siting process is defensible before regulatory agencies and the public.
- **Differentiation.** OSI employs the Opti-Site site selection process. This process was designed based upon decades of siting experience and it utilizes the axioms of decision analysis to optimize the siting process, and integrates decision analysis into a geographic

information system (GIS) to create a powerful decision model for site selection.
- **Proof.** OSI has successfully identified four similar solar projects in California that were sold to developers and are now operating facilities. In addition, OSI performed site selection for major industries here and abroad.
- **Tangible Results.** OSI's site selection process identifies sites for solar energy facilities that have a higher probability for success in winning a Power Purchase Agreement from the local utility company. That is, the OSI siting process results in lower business risks.

How Do You Know Your Value Proposition is Strong?

Tim Williams, author of *Positioning for Professionals: How Professional Knowledge Firms Can Differentiate Their Way to Success* (Williams, 2010) offers several questions that can be used to test your value proposition to ensure it is strong:

- Does your value proposition feel authentic?
- Does it make your firm's and your offering intensely appealing?
- Does it have a strong barrier of entry (i.e. difficult for competitors to copy)?
- Is it difficult to find an exact substitute?
- Does it result in fewer competitors?
- Have you created a truly customized offering?

Why and Where to Use a Value Proposition

Why do you want a value proposition? An important reason might be surprising: to help you understand your service offering! That's right. Have you really sat down and thought about your offering (s) in terms of congruence, benefits, differentiation, proof, and tangible results? Use this value proposition template to *improve* the value of your offering. This is a worthy exercise.

Where should you use a value proposition? Use your value proposition in all of your marketing materials, including your sales letters, proposals, and sales presentations.

Finally, you'll find that your value proposition is especially helpful in creating your elevator speech, your one-minute description of what you do that comes in handy in conferences, or other places where you meet potential clients who want to know what you do.

Give a prospect what he or she wants and take the price out of the buying decision!

Gaining the Edge with Marketing Psychology

When we are marketing, we always want to gain an edge on the competition and convince the prospect to buy from us. In other words, we want to *persuade* the prospect to buy from us. Marketing psychology is all about persuasion. Marketing psychology is another arrow in your marketing quiver.

Several giants in the field of psychology and marketing have made real contributions to persuasion and influence, and these concepts have proven to be quite useful and effective. We'll cover three "experts" in persuasion to begin this chapter.

A History of Persuasion: Aristotle

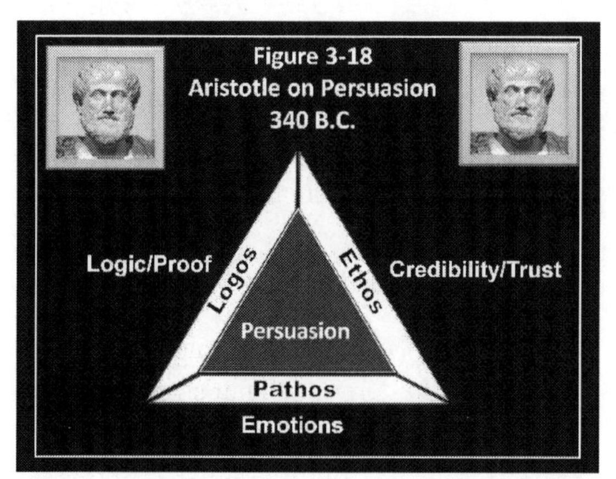

Aristotle, around 340 B.C., got it. He nailed the theory of persuasion. Most writing about persuasion since then mirrors this theory. He described the nature of persuasion in terms of three factors: ethos, pathos, and logos, as shown in Figure 3-18.

Logos represents logical justification or proof, as expressed by case histories, comparisons of alternative solutions, statistics, and facts. Logos is a logical appeal. Logos makes the consultants appear knowledgeable.

Ethos represents an appeal to authority or credibility of the consultant. Ethos shows the consultant is qualified to present an offer or suggest a solution.

Credibility can be demonstrated through a personal brand, a track record, testimonials, and successes or awards.

Pathos is an appeal to the prospect's or client's emotions. Pathos can appeal to pain or fear, as well as pleasure or compliance. Pathos can appeal to the prospect's imagination, hopes, and feelings. Emotions come through stories and vivid language.

A History of Persuasion: E.K. Strong

In 1925, E. K. Strong proposed AIDA, an acronym for a persuasive sales process based upon how, according to Strong, a person decides to buy. This acronym is repeated throughout marketing literature today. AIDA is as relevant today as it was in 1925. It mimics how a buyer moves through the buying process.

Here is what it means:

- **A=Attention:** Grab the prospect's attention quickly. Ask a question like "Have you ever….?" "What would happen if you…."
- **I=Interest:** Once you get their attention, *make a bold promise*, or tell them something about how you will solve a problem they have, or get them to open up about their problem. "We'll get the job done in half the time".
- **D=Desire:** *Educate the prospect* about your service and tell them how you will solve their problem.

  Emphasize benefits. "Just how will you get the job done in half the time?"
- **A=Action:** Ask for the sale, or the next move (e.g. "Should we meet Tuesday or Wednesday next week?").

A buyer goes through a multi-step process. If you ignore a step, or change the sequence, sales effectiveness drops.

There are two learning points here: that *you need to get attention first, then create interest, then build desire*, and finally encourage an action. And second, if you leave out one of these factors, the effectiveness of your sales message drops *significantly*.

A History of Persuasion: Eugene Schwartz

The next person who impacted marketing psychology was Eugene Schwartz with his book, *Breakthrough Advertising* (Schwartz, 2006). Schwartz introduced four concepts that impact persuasion:

First, Schwartz said an offer can be strengthened by the content of the offer, the sequence of subject or topic presentation, and prior positioning or preparation.

Second, Schwartz describes three dimensions of feeling and thought that affect persuasion: desire for the offering, conviction, and by that he means the need for expression, longings for social status, hopes, dreams, ambitions, goals, recognition, mastery, admiration of associates, prestige, all of which will be unspoken and unstated, and beliefs, the filter through which a prospect sees your offering. *Schwartz made an important point that you must address the prospect's covert, or unstated, concerns and desires.*

Third, The Rule of Beliefs: if you violate a prospect's beliefs, you lose. Rather, you must channel the force of his beliefs behind your offer. This is a powerful concept.

Fourth, Schwartz calls it "concentration", the process of pointing out weaknesses to approaches used by the competition, and then proving to the prospect that your offering gives him what he wants without these weaknesses. In other words, demonstrate that other solutions will not work by showing options, and eliminating options likely used by your competition.

A History of Persuasion: Robert Cialdini

Robert Cialdini has written several books on persuasion and influence, including *The Psychology of Persuasion* (Cialdini, 1984). Cialdini performed research into the factors, or principles, or psychological cues, or triggers, that influence or direct personal behavior to bring about compliance with a request.

Cialdini characterizes these principles, also called judgment heuristics or mental shortcuts, as weapons of influence. These mental shortcuts enable us to make quick decisions based upon a minimum of information.

An example of a judgment heuristic: if it is expensive, it is good, or, if an expert said it, it must be true. Even, if a celebrity said it, it must be true! We'll look at a number of these judgment heuristics in the later in this chapter.

The Elements of Persuasion

Why do your prospects buy from you? That is a very important question that all consultants must think about and be clear about. Answers to this question will drive you in a certain direction. Is it the right direction? Let's look at a few possible answers.

Does the prospect buy because we're the lowest cost?
Does the prospect buy because we offer the best service?
Does the prospect buy because we offer the highest value?
Does the prospect buy because we offer the best solution?
Does the prospect buy because our reputation is superior?
The correct answer is....

Because they want it!

Your prospect will not care about the questions on this list unless they want your service offering....in other words, they become emotionally involved in wanting it.

These questions relate to logical justification. Your prospect will buy from you only when they can sense doing so will enhance their self-image through recognition, learning, pride, reduction of anxiety, enjoyment, etc. Other ways to get the prospect emotionally involved include greater promises, more dramatic images or visualizations, and more compelling proofs.

The psychology of marketing is all about persuasion. What is persuasion? Here's the best definition I have seen:

"Persuasion is the conversion of unformulated desire into high resolution images of fulfillment."

In other words, a prospect initially has a fuzzy desire or an unformulated desire, or a general desire for something. A consultant's job is to make that desire much more specific, more intense, and focused on their offering.

Turning this definition into an equation, and adding another term, we have:

Persuasion = Believability + Aspiration + Desire + Logical Justification + Auto Response

So, we're saying that to persuade, we must get a prospect to believe in what we are selling, we must appeal to a prospect's self-interest or self-image or longing for social status, also his dreams, his hopes, collectively called his aspirations (i.e. covert wants), we must satisfy the prospect's desires for tangible things (e.g. a solution to a problem, or overt wants), and we must provide some logical justification for the purchase. The auto response term is explained below.

Let's do a deeper dive into the five terms of the Persuasion Equation:

- **Believability.** Your proposal is filtered by your prospect's beliefs or convictions, as depicted in Figure 3-19. Those beliefs consist of opinions, attitudes, prejudices, fragments of knowledge, feelings, and past experience. As we mentioned earlier, one's beliefs are a source of emotional

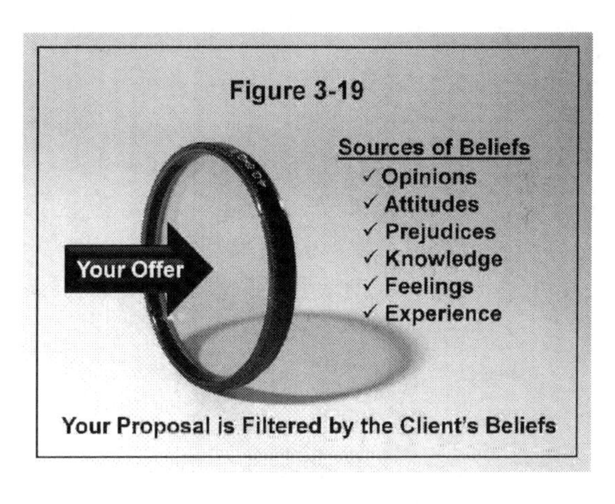

Your Proposal is Filtered by the Client's Beliefs

security and certainty. Think what happened to Galileo, who championed the heliocentric world rather than the geocentric world that was believed for thousands of years. He was convicted of heresy and committed to house arrest for life. This story illustrates the power of beliefs, and especially conviction. Conviction eclipses a belief because it includes a powerful, intense emotion. The key here is that by the time you write your proposal, your prospect's beliefs are hardened, and it is very likely they cannot be changed by words. Rather, we must channel our proposal messaging to be consistent with the prospect's beliefs and convictions. Then our message will *resonate, not conflict.* We will discuss how you might change a prospect's beliefs later in the next section of this chapter.

- **Aspiration.** The prospect may aspire to being recognized by his superiors, higher social status, more prestige, being admired, fulfilling ambitions, inclusion in a select group, achieving personal goals, or achieving professional goals. These are unstated, or covert wants, which are more difficult to learn. Nonetheless, they are critically important.
- **Desire.** Desire addresses more material wants such as losing weight, selecting the right consultant, meeting a budget or schedule commitment, finding a solution to a problem, buying a new car, purchasing new field equipment for the office, higher profit, increased revenue, or reduced risk. One purpose of a proposal is to sharpen desire.
- **Logical Justification.** When a prospect is considering a purchase of a service from a consultant, the purchase decision is based on both emotion and logic, as we have seen. In every purchase decision, there is uncertainty. The role of logical justification is to reduce this uncertainty to a point where the purchase can proceed. Another purpose of logical justification is to tell the reasons why the prospect should buy. Reasons include what the service does in detail, how it solves the prospect's problem, and, of course, the key benefits for the prospect.
- **Auto Response.** Auto response is a reflex action in response to a mental shortcut or trigger. Making purchase decisions is quite complex these days. Our prospects must act quickly, with less

information and data than they would prefer, and choose the best solution amongst many attractive-looking alternatives. When their mind is overloaded, they tend to create shortcuts in their thinking and judgments. They will look for an easy way out of the complex decision. In so doing, they look for indicators, or signs, or signals that can reduce the complexity of the decision, to simplify it and provide some comfort that they are making the correct decision. As a result, they may *narrow* their attention and rely on a *single* piece of information, or trigger, that moves them toward a decision, almost like an automatic response. For example, Amazon always tells you how many of items are left because scarcity often pushes us to a decision. These triggers are also called "psychological cues" or "judgment heuristics". We discuss triggers again shortly.

These five terms of the Persuasion Equation have a real impact on any decision, and are your toolbox for persuasion. We discuss how to do this in the next two sections of this chapter.

Applying Marketing Psychology I

In this section, we discuss how to apply the first three terms of the Persuasion Equation to the marketing process. The goal is to augment persuasion by increasing believability, desire, and aspiration for your offering.

- **Increasing Believability.** Sure, we want to increase the prospect's belief in our offering. However, this is very challenging. As we mentioned earlier, people's beliefs bring with them a strong sense of certainty, even security, about how things work. This feeling of certainty can help us take action. However, this certainty also can hinder, or even block any changes to their beliefs, especially if the beliefs are convictions. At the proposal stage, all we have left to change beliefs is words and images, which are much less effective than face-to-face conversations. If our offering is inconsistent with the prospect's beliefs, the best we can do is address what we think are the objections the prospect has regarding the offer. However, if our offer is consistent with the prospect's conviction, even an average offer is likely to be well received. That is how powerful an offer can be if it is congruent with belief.

- **Increasing Desire.** This term in the Persuasion Equation is much easier to maximize than belief. Here are a few ways to increase desire in a proposal or presentation:
    - A compelling sales message, derived from a strongly crafted value proposition;
    - A strong element of proof (i.e. a relevant case history or testimonial);
    - Education, which results in a higher level of specificity about your offer, and shows the prospect that you are an expert. If you have a better way to solve a problem, describe it in detail.
    - Showing the prospect what the future looks like with your service implemented. This one is especially powerful. For example, when I prepare a proposal for a site selection project in the power industry, I show a simulation of the completed project on the proposal cover.
    - Comparing your offer to an inferior offer, or comparing your offer price to a costlier offer (i.e. use options to present your offer in a better light).
    - Add a guarantee. Guarantees are a very powerful reducer of emotional risk. Look for multiple ways to reinforce the level of desire in your proposals and presentations.

- **Increasing Aspiration.** Recall that by aspiration we mean, for example, the need for recognition, prestige, social status, a promotion, looking good to superiors, learn something new, or satisfy ambition. Also, the prospect may want to reduce his emotional risk, which can be accomplished by increasing his sense of certainty. These are covert, or unstated needs that should be addressed. The only way you can address these concerns is to build a good relationship with the prospect *before the proposal is issued.*

Your responsibility as a marketer is to look for ways to increase believability, desire, and aspiration, i.e. do the best you can to persuade the prospect to buy from you. Persuasion is an important tool.

Applying Marketing Psychology II

Earlier in the chapter we talked about auto response, or psychological triggers or cues. A psychological trigger induces desire via either reducing anxiety or increasing buyer comfort). These triggers occur below the level of conscious awareness, and are alternative paths to decisions or actions. Robert Cialdini describes a number of psychological cues in his book *Influence* (Cialdini, 2006) and Joseph Sugarman mentions additional cues in his book *Triggers: 30 Sales Tools You Can Use to Control the Mind of Your Prospect to Motivate, Influence, and Persuade* (Sugarman, 1999). We provide a few of these cues that are particularly useful for consultants who write proposals or give presentations.

- **Scarcity and Urgency.** These are well-known triggers for action. "Only 100 copies were printed so order you copy today" is an example of the use of scarcity. When discussing an upcoming opportunity with a prospect, indicate that your best Project Manager for the job is available now, but you cannot guarantee availability in the future. Urgency creates desire. For example, an upcoming regulatory deadline creates urgency and encourages action. Look at Ebay auctions. As the time period closes, you see urgency and scarcity in action!
- **Reciprocity.** This also is a well-known trigger, and one of the most useful for consultants. Give your prospect or client something of value. This could be a free report, a gift, or some free consulting, such as an audit. Giving stuff away *also creates an obligation to return the favor.* Giving a prospect something for *free creates a feeling of indebtedness*, and most of us don't want to be in debt. Other examples include breakfast, lunch or dinner with a client, or tickets to her favorite sporting event.
- **Loss Aversion.** Loss aversion is important psychological process that greatly affects our buying decisions. Many people who own stocks like to sell stocks that have gone up, thinking that they are taking a profit. This leaves only stocks that don't go up in their

portfolio. This is an example of wanting an immediate reward, but not valuing the fact that the stocks that went up are the best stocks in the portfolio. Immediate rewards are overvalued in comparison to what will happen in the future. We seek instant gratification rather than losing that opportunity. How can you use loss aversion in marketing? *Point out the dangers along the way to a solution.* What will happen if they don't solve the problem? Does that create frustration or worry? In my siting practice, I often ask the question *"What would happen if the site you chose turned out to have a fatal flaw?"* Each of the attendees now imagine a terrible scenario in their minds. I quickly add that our siting process virtually eliminates this scenario. Finally, ask questions that start with the following phrase. What would happen if [blank]. "What would happen if you go forward with the acquisition without evaluating the environmental remediation risk"? Tell your client about the risks of no action or the wrong action.

- **Visualization.** The concept of visualization, or anticipation, is another marketing technique with psychological value. Anticipation is a key stage in happiness. Tell an audience that you have great news later in your presentation, and they will stick around to hear about it. Publicity builds anticipation. You've probably found yourself tearing while watching an emotional scene in a movie. The movie isn't real, but your mind reacts as if *it is* real. How do you utilize the benefits of visualization in consulting? First, what will success look like when your solution is implemented? In other words, describe the future to the prospect with you or your solution in it. You can do this with words or better yet, with images or simulations. For example, when we are communicating to the public about developing a large ground mounted solar facility, we get questions about the visual impacts. We answer those questions by showing a simulation of the

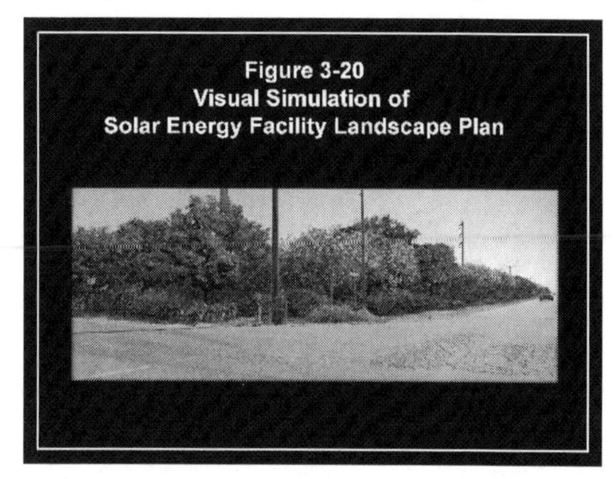
Figure 3-20
Visual Simulation of
Solar Energy Facility Landscape Plan

landscaping. An example is shown in Figure 3-20. The solar energy project is hidden behind the landscaping.

- **Authority**. The concept of authority is a well-known influence trigger. You've probably used this one! *Authority is always perceived as powerful.* There is a strong willingness of adults to go to almost any lengths on command of an authority. *Someone with authority is not questioned! Do you question your doctor or lawyer?* Titles also are powerful symbols, like President, Director, PhD, P.E., Dr. Being a recognized expert is an excellent trigger. How can you demonstrate your authority? Describe your expertise. Mention your certifications or licenses.
- **Comparison.** The concept of comparison, or contrast, is important to understand. For example, get the prospect to buy a small service to start. This purchase will change the prospect's view of you, and they will be much more open to a larger upsell. I've used this technique many times with compliance audits or environmental assessments. These assignments often lead to larger jobs. In the site selection business, I will sell a low-cost task called a Critical Issues Assessment for a proposed site. If the assessment indicates an attractive site, I'm well positioned to conduct the full environmental studies of the site. Compare your offering price with a higher price first, then show your price (lower). After seeing the higher priced item, a less expensive item will appear even smaller in comparison. Offer a volume discount. Or, explain that if field work is conducted, your service would cost $40,000. However, you already have sufficient data, saving $15,000, so the price is only $25,000. Finally, indicate that a full-blown Environmental Impact Report will cost $200,000, but you believe that a Mitigated Negative Declaration will be approved, costing only $50,000 to complete.
- **Third Party Data.** The concept of third party data is very helpful in further convincing your prospect that your argument is valid. This is similar to using authority. Also, third party facts increase *credibility*. For example, if you are offering a fire prevention course, provide the prospect with statistics on fire damage. I was presenting a solar energy project to a Planning Commission. There was controversy surrounding whether the soils at the site qualifies as prime farmland. We engaged the most respected ag soils lab in the county to sample the soils and provide their professional opinion: that the soils did not meet prime farmland standards. The independent lab data lended credibility to our position. A classic use of third party data was a

carpet cleaner by the name of Joe Polish, who found a government report showing that carpets are often homes for a number of nasty bugs. When housewives saw this, they got the cleaning job every time! Think about how you can use third party data that will support your arguments.

- **Experiential Involvement.** This cue applies primarily to presentations (formal and informal). The idea is to get your audience involved in the presentation or meeting. In other words, you need to give the audience an experience that relates to your service. There are two ways you can do this: *a physical way, and a thought-provoking way.* The physical way is to use some props that put the audience closer to your service. You might show them some equipment or photographs that demonstrate how your service or product works. For example, we were selling rooftop solar panels adhered to a roof using industrial grade Velcro. We would demonstrate it in a meeting by bringing a panel attached to roofing material. We invited the guests to try to remove the panel. They couldn't. The thought-provoking way is to make your audience get involved mentally in your presentation. Give the presentation like telling a story, where the audience imagines your experience. Don't tell them the ending first and destroy interest.
- **Instant Gratification.** The concept of instant gratification is a well-known influence trigger on the Internet. For example, most offers give you instant access. Whatever you can buy, it will be delivered within *seconds*. Buyers love this. When they decide they want something, they want it *now*. Your prospects and clients are no different. Offer to begin work without the contract in place (not a lot of the work of course!). Do work in a proposal to show "instant progress". In my site selection practice, I often include a GIS map in the proposal (24"x36" -- remember experiential involvement?) that shows a number of data layers, with some insights noted. The map gives the impression that I'm already working on the project. Think about how you can market your service in a way that you will respond to the needs of the customer virtually immediately.
- **Control.** The concept of control can play an important role in marketing. Why is control important? *A client always wants to be in control.* If they think they aren't in control, they see a risk in buying. How can you increase the prospect's control? One way is to break down the job into smaller phases, and just ask for authorization of the first phase. This increases the client's control over the budget. Second, I always break down my site selection assignments into 4-6

tasks. I provide the option to authorize just one task at a time. Even if the client does authorize the entire job, I go to great lengths at the end of each task to ensure that my client is fully on board with the results and what we will do next. Third, try putting the project manager or principal investigator in the client's office, periodically or full-time until the job is well underway, or even completed. The client will feel more in control because his consultant is right down the hall. Fourth, propose frequent status meetings or work meetings with the client. Meetings provide opportunities for clients to exert control. Fifth, provide written progress reports each week with a follow-up call with your client to answer any questions. Think about how you can give your clients more control.

Now you have a number of methods that will increase your ability to persuade a prospect to buy from you. *The more of these triggers you can apply in a presentation, proposal, or sales letter, the more convincing you will be.*

Influence Magnets

We finish this chapter with a few influencers, or influence magnets, that clients will appreciate, and, hopefully, will return the favor.

- **Make Your Client Feel Important.** Do something to make your client feel important. In one instance, we were working on a project. We had the lead since our expertise was needed. However, the client was located in Texas, and I was in California. So, I convinced my client contact to assume the position of Team Leader, since some of the work activities could be conducted by the client's staff. By doing this, I put my contact in a leadership position which gave him some good visibility with his superiors. Another example is to write a paper for presentation at a conference and include your client as an author. The client gets recognition and a nice trip to the conference location.
- **Teach Your Client.** Everyone loves to learn. Find a topic that your client would like to know more about and arrange to teach the topic. Start with business-related topics, such as technology, new regulations, or new software. Then see if the client is looking to learn more about something *personal.* For example, I had a client who played very average golf. I invited him to a golf club that my father belonged to at the time. We started by hitting a few balls at the

driving range. I arranged for the club pro to *happen by*, and he gave some pointers to my client. He was absolutely thrilled.

- **Best Practices.** Clients like to hear about best practices in their industry. This is always a great way to engender respect from a client provide them with the latest developments, or methods, or practices that will help *them* succeed. I'm always trolling the trade organizations in my client's field of interest. When I see a new report that might be of value to my clients, I send a short email with the link to them. Keeps me in their minds when the next opportunity arises.

You are now armed with a selection of persuasion and influence techniques. Create a table listing all of them and keep it at your desk. Review them as you prepare proposals and presentations. You will be rewarded.

Positioning: Do This First!

"Positioning means your prospect is pre-disposed, pre-interested, pre-qualified, and pre-motivated".

Joe Polish

Positioning in the context of selling services means to create a "position" in the mind of your prospect or client that reflects the strengths of your firm, the benefits of your service offering, and what abilities you bring to the situation. The idea is to position yourself to the prospect before you meet to discuss an opportunity, before you make an offer, or before you begin the selling process in earnest. The advantage is that when you do meet, you can go directly into discussing the opportunities rather than using valuable time to cover background material.

The goal is to make the prospect more receptive to your subsequent offer. What you don't want to do as a first step is submit a basic information brochure that just presents your office locations and basic services you offer. Rather, you must be much more detailed and specific to the nature of the prospect's needs and your offer. In other words, create a "proto-proposal" or preliminary proposal that emphasizes the following:

- Key benefits based upon a preliminary value proposition;

- Use of emotion by addressing the prospect's underlying frustration; concern, or fear regarding a problem, or addressing the prospect's aspiration;
- Exude confidence, optimism, and certainty;
- Demonstrate committed interest and care about the prospect's situation;
- Display authority and credibility;
- Educate the prospect; and
- Describe a case history similar to the prospect's need, and how you won that job.

Positioning involves four perspectives: the corporate perspective, the offering perspective, the personal perspective, and the prospect's perspective.

Corporate positioning includes longevity, qualifications, and experience, all geared towards making *your firm* different.

Offering positioning focuses on the value proposition and why *your offering is unique or different.*

Personal positioning addresses how *you* are different, including your communication advantages, who you are (e.g. your credibility, authority), and your *specialized* expertise.

Prospect positioning involves showing that you are quite familiar with the prospect's firm. Research data on the firm. Read the latest financial reports. What are their goals? How are they organized? Search LinkedIn to see if you can connect with any key personnel. What challenges do they face? What regulations must they comply with? Where are their offices/plants located? Are there any articles about the firm in trade journals?

One of the best ways I've ever seen to position is to create a newsletter or email campaign that provides valuable information to the prospect over several months. I've seen this approach not only win jobs, but create entire business lines.

IV. Key Skill No. 3: Client Service Strategies

Now we move away from business development issues and introduce Key Skill #3. In this chapter, we address a variety of topics relating to working with your clients, all with the aim of improving your relationships.

What Clients Want from Their Consultants

We begin this chapter with a very basic question: *what do your clients want from you?* We present a dozen examples of what clients want from their consultants. This list is the result of my several decades of consulting.

- **Change.** Clients want change. That's why they need help. That's why they want consultants. You must show how *you* will bring change. What do we mean by change? Change means improvement over the current state. How will you change the current situation and improve the client's results? How can you deliver change to your prospects and clients?
- **Perspective.** Your client wants to know what his/her peers are doing and what agency regulations are on the horizon. *They want to know the "big picture"*. Once again, do your research and be aware of what is going on in the client's market. What are his competitors or peers doing that they should be doing? How can you bring perspective to your prospects and clients?
- **Cutting Edge.** Clients want the cutting-edge technology or the latest ideas. This is critically important. Clients expect that *you* will have the latest and greatest. If you don't have a new technology, package your service in a way that *infers* that you are at the cutting edge of your field. Research the latest innovations in your field, and include at least one new idea, and mention this in your marketing. Being at the cutting edge always has been the foundation of my marketing

message regarding my site selection process. *What can you bring to your clients that is cutting edge?* Hint: cutting edge doesn't need to be related to technology!

- **Functionality.** Show the client how your solution will work or how it will be used. They want to know *how* it is going to solve their problem. *Functionality also means flexibility.* Show that your approach to providing a service is flexible, so that you can accommodate unanticipated events, or apply the service under varying circumstances or conditions. How can you demonstrate functionality for your service offering?
- **Energy.** Clients want to see self-confidence, enthusiasm, and a positive attitude. They want to see that you care about them and that you will take the extra mile to help. They want to see that you have the confidence in your ability to solve their problem. Be prepared to show why you are confident that you can help.
- **Brand.** You need a brand that the client will recognize. For example, I am recognized as a siting expert, others are recognized as regulatory experts, wetlands experts, hydrogeology experts, and so on. Position yourself to be a brand. Brand all of your marketing materials and website. Try writing a personal brand statement and use it in your marketing material. This statement should be a short paragraph in length and address what attribute(s) sets you apart, what audience you serve, what you do, and why you do it. Your brand is a powerful tool that can increase trust and credibility. Try adding your brand statement to your resume! That surely will set you apart from your competition.
- **One-Stop.** Clients want one-stop shopping, all-in-one, and would rather work with one consultant than several to solve what they see as one problem. That doesn't mean you have to be an expert in everything! If needed, create a team to solve the client's problem. Bring in other consultants or contractors that you know or who are recommended by people you know. I have been in this situation often. I get a potential client interested in my service, but then the client adds other tasks to the scope. I go out and get other experts to address these issues for me.
- **Commonality.** Your client wants to see that you understand their situation, speak their lingo. How does this happen? You must get to

know your client from a personal perspective. When you first enter a client's office, look around. What is hanging on the wall? What pictures are on the desk? What awards are framed? Look for something to comment on and create some commonality. I was introduced to a prospect awhile back, and I thought it would be difficult getting to know him because his interests, and cultural background, were so different than mine. However, he mentioned that he was part of a new business enterprise many years ago. It turned out that I was as well! From there on, our relationship bloomed.

- **A Framework.** Frameworks show a deep understanding of the problem/subject and a path to a solution. Clients like to see that you can frame their problem and compare it to other problems you or others have solved. And, of course, frameworks help you understand a problem or situation. One of the first things I do when selling site selection services is to pull out my site selection framework. It becomes part of your brand. How can frameworks help your practice?
- **Value Pricing.** Clients want to know they are getting good value, even if your billing rates are high. First, you can demonstrate value by telling the prospect exactly how your service delivers value or second, use comparison pricing to put your offer in perspective. For example, compare the savings your service will create for the prospect to your fee. Comparing a large savings to a much smaller fee is a strong argument. When I offer due diligence services, I always compare my fee to the potential risk costs that can be avoided. How would your service increase your client's profit or put them at an advantage to their peers or competitors?
- **Show the Future**. Clients want to know what the future looks like when your solution is implemented. Consultants often focus only on their solution, how and why it works. Take another step and describe what your client's future will look like *after* your solution is implemented. *Clients are impressed when you predict the future in a confident manner.* In my site selection practice, I often create a proposal cover showing a proposed site with a new facility, thus showing the client what his future would look like…. his fully developed project!

- **Leadership.** Clients are looking to be led. What does it mean to be led by a consultant? First, a consultant-leader stands up for what he believes is right for the client. Always. Second, *he isn't afraid to challenge the status quo* and suggest new ideas. Third, he is willing to recognize other good ideas, especially the client's good ideas. Fourth, a consultant-leader always looks to the future by showing how the client can succeed in the future. Lastly, a leader has that perspective of the business we talked about a moment ago.

So, once again, take time to examine each of these twelve "client wants". Every one of them is important, and if you specifically address them, you will be surprised at the reception you get from your prospects and clients!

What a Consultant Needs to Know About Their Client

One of the most important tasks for a consultant is to learn the needs of the prospect or client. Over my career as a consultant, most client needs have fallen into one or more of twelve needs. Now, these are what you could call unstated or implicit needs. Understanding which of these needs applies to your prospects is important in order to properly respond to a prospect's wants and needs, and build the relationship. These are general categories of needs, of course. In a given situation, the needs will be spelled out in more detail, but likely will fall into the ones mentioned here.

- **Work Quality.** Clients differ on how sensitive they are to work quality. All clients want quality, but some want extreme quality. You should determine that sensitivity early on. An indicator is how the client looks at a draft report. If they go crazy over a typo in a draft report, that's a sure signal they want total work quality. Make your draft reports look like final reports.
- **Responsiveness.** Once again, some clients look at responsiveness as the true test of commitment. Responsiveness means how fast will you return a voicemail or email inquiry. Responsiveness also means how quickly you can conduct a study or write a report, or go from draft to final report. *Some clients are poor planners and expect their consultants to jump through hoops to make up for their shortcomings.* If responsiveness is a hot button, install an internal procedure that will ensure quick response to this client's needs.

- **Problem Understanding.** Are you and your client on the same wavelength when it comes to understanding what the client wants? Are you listening to the client to determine their needs, or are you assuming what those needs are? Most clients want to go through a period of self-discovery when evaluating a problem. They appreciate your help along the way, but they don't want you to tell them what they need. When they discover it, then they will tell you.
- **Project Management**. Some clients really want someone to manage their work because they don't have anyone internally who can do it effectively. If this is the case, making an offer to work more closely with this client, and even station a project manager in the client's office for a time, may be a hot button for them. How sensitive is your client to selection of a Project Manager? See High Maintenance below.
- **Price.** Some clients are focused on price first. Now you shouldn't have many of these clients anyhow. However, that's probably an oversimplification. If you have a good client that happens to be sensitive to price, you must look for creative ways to work within that constraint. Spend time finding ways to be more efficient, and spend their dollars carefully, and let them know that's what you do. Give them contractual control. Provide free stuff once in a while.
- **High Maintenance.** These clients like it their way. Give them a report to review and they practically re-write it for you. Then they complain that they had to do this. And the real meaning of the report didn't change. Just the words. These clients are detail oriented and get stuck in the weeds. They are apt to call at any hour looking for help. *Everything is urgent. They can drive you crazy.* These clients need project managers who are very patient. I had a client like this a while back. I would take the client out for golf, fishing, lunch and build up a bank account of goodwill that I could spend when the client was not happy with something we did.
- **Access.** Some clients want instant access to your senior managers. Whenever they have even the smallest beef, they want to talk to the boss. *These clients must have a senior executive assigned to them*, and they must frequently contact the client to see how they are doing. This will reduce the times the client will insist to talk to senior management.

- **Problem Solving.** Some clients are analytic in nature and enjoy solving problems. In this case, *your project manager should be a technical person* who also enjoys getting into the details of problem solving. These clients are creative and like to learn new techniques.
- **Security or Safety.** *These clients are hypersensitive about failure. They are worriers.* They are probably not familiar with the details of your assignment. *They need hand-holding.* Or, they are very sensitive about health and safety procedures. In this case, be sure that your staff who work in the field for this client are well versed in health and safety procedures. Their health & safety plans are up to date and complete, and their field preparation is flawless.
- **Relationship.** These clients thrive on a close relationship with their consultant. They like to be entertained, via lunches, dinners, golf, fishing, etc. These clients need a project manager who enjoys the same, and will take the time out of their life to spend with the client.
- **Knowledge or Expertise.** Some clients always want the A Team. They want staff with several letters after their name. In this instance, load up your project organization with those personnel. Make sure they are involved, but if you can, limit their involvement to scoping and review to control cost.
- **Dodgy Client.** What I mean here is "untrustworthiness". For example, this client will not say a word without signing a Non-Disclosure Agreement. And, it is one-sided -- his side. Then you get to the contract. This client always seems to have a problem. This client thinks you will rip them off. Stay clear.

It is difficult to typecast clients. However, keeping these types of needs in mind has helped me deal with them more effectively, and with lower stress. It is always better to understand your client's idiosyncrasies.

Why Clients Don't Buy from You

There is a lot to be learned by understanding why clients *don't* buy from you. The following reasons were the most frequent ones I saw in my long career in consulting:

- **Client Doesn't Understand the Service.** When I first started in consulting, I often lost opportunities for this reason. I finally realized that I didn't clearly and concisely explain what the service or offering was and how it worked. For example, I once had an opportunity to conduct a due diligence study of a 25-acre parcel that a developer wanted to purchase for commercial warehousing. There is a standard process for this type of work that is published. So, I just copied the steps in the standard, and slapped in a cost. I lost the job. Why? My competitor more fully described his scope, which included a number of soil samples to be taken from potentially suspect locations. I described the "what" (a due diligence study) but I didn't describe the "how" and the "why".

- **Client Doesn't Need the Service.** In some cases, this is code for *"I like the consultant I use now"*. Or, perhaps you haven't sufficiently described the "what" and "how" your service can help them.
- **Client Doesn't Trust You.** This response typically means that you didn't spend sufficient effort describing the offering, demonstrating the benefits, *proving credibility*, or building the relationship.
- **Client Cannot Obtain Approval.** This response typically occurs when you talk to someone who doesn't have the final authority to hire you. Be sure you know who the real decision-maker is. Or, this response may be code for "I am happy with my consultant."
- **Client Cannot Justify the Cost.** Over the past ten years or more, I can't remember any instance where a prospect or client told me they couldn't afford it. Either of two situations will precipitate this response. Either you are not adding sufficient value to attract the buyer, or you didn't differentiate your offering very well from the competition. When you provide tons of value, cost is never a problem.

The takeaway here is to think ahead about the reasons why you think a prospect will not buy from you, and be prepared to overcome these objections. Of course, the best way to deal with objections is to ask the prospect about them before you submit a proposal.

Questions to Ask Your Client at Kick-off Meetings

A kick-off meeting for a new project can be the most important meeting the project will have on that project. The kick-off meeting is the only opportunity to achieve a meeting of the minds between the client and the consultant, while the project is still, what we could call malleable or in the honeymoon period. You must take advantage of this situation to get as clear as possible about the needs of your client on the project.

Here are a few basics you should cover at the kick-off meeting:

- **Communication.** Set up communication protocols for the project. Get phone numbers and email addresses. Can you call the client after hours? What are the best times of the day to reach the client? If your primary contact is unavailable, who can you reach? Does the client have any confidentiality issues?
- **Concerns and Objectives.** Review the project background and objectives. What are the key project concerns? How should they be handled? Do their answers affect the scope, budget, or schedule? What are the client's expectations regarding project outcomes? How will project success be measured?
- **Submission Preferences.** What report format is preferred by the client? How about style issues? What is their approach to draft reports? Do they prefer "perfect" drafts? How long will they require to review drafts?
- **Update Preferences.** What are the client's preferences regarding progress reports or progress updates? Does the client want a written progress or an email report or an oral report by phone? How often does the client want these updates? What is the format for progress reports?
- **Budget Allocation.** Is the budget allocation consistent with the contract amount? How is the budget to be expended? If the budget is broken down by task, and you overrun one task, can you borrow from another task budget? What format is desired by the client for invoicing? Who is the client contact in their Accounts Payable group?
- **Project Schedule.** Is there agreement on the schedule or milestones? Is there any contingency in the schedule?
- **Changes.** How will scope change requests be handled?
- **Team.** Who will be involved on your team? Are there minority contracting or subcontracting requirements? How should subcontracts and tracking of subcontracts be handled?

- **Reviews.** Does the client want quarterly project reviews or other more formal project reviews?

The project kick-off meeting is the best way for a "meeting of the minds" between the consultant and client to be achieved. Don't miss this opportunity. Also, a kick-off meeting is a great time to bond with the client team. To this end, it is helpful to end the kick-off meeting with lunch or the like.

The Client Feedback Process

From my experience, the Client Feedback Process is the most important opportunity for consultants to build relationships and market their services to existing clients. There is a tremendous amount of value to be harvested from interviewing clients regarding your performance.

Here is a three-step approach to the client feedback process

- **Step 1:** Before carrying out a client performance interview, you should assemble a list of questions you want to ask. This is absolutely critical if you want to maximize the value of the feedback event.
- **Step 2:** Select the person to do the client interview session. This is important. Options include the project manager, the senior executive on the job, and an independent senior level person not associated with the project. From my experience, the second and third options are best. I like using an independent interviewer not involved on the project. They tend to obtain the most unvarnished replies. Obviously, if you are in business for yourself, you conduct the interview! By the way, *always do the interview at the client's office.*
- **Step 3:** After the interview, it is the responsibility of the project team to respond to the feedback. There is nothing worse than ignoring the client's comments!

Here are a few questions that I have used many times in client feedback sessions:

- **"What do you like best about working with us?"** A good opening question that should get you off to a positive start, which is important.
- **"What is the most disappointing experience you've had with us in the past year?"** Ask this later in the interview after a few positive questions. Don't ask this question directly after the first one!
- **"What can we change about your relationship with us to make it more effective for you?"** A great open-ended question that almost always yields helpful feedback.
- **"If you have a problem with our firm, do you know who to contact?"** It is very important for the client to know who to contact if they have a problem with your firm.
- **"Are our rates in line with the value we provide?"** This is a super question! In some instances, clients will hesitate to provide a clear answer to this one, but it's worth asking.
- **"During the past year, do you feel that your dealings with us have improved, stayed the same, or deteriorated?"** Another very effective question. After the response, ask why!
- **"Are there any other issues that you want to discuss about our performance?"** This is a nice, open-ended question that fits at the end of your interview.
- **"Where do you see the market in the next year or so?"** This question should be used at the end of the interview if things went well. There is nothing wrong with probing marketing opportunities at this point in the interview. This is an essential question to learn about the client's future plans.
- **"What opportunities do you see for us in the future?"** A good follow-up question to the previous one. Probe a bit deeper.
- **"When buying consulting services, what are the biggest mistakes consultants make when working for your firm**?" This question tells you what NOT to do!
- **"Are we communicating effectively with your team?"** This is an important question that often gets an unexpected response.

- **"Do you feel that we are devoting adequate resources to your project?"** This is another question that elicits a response almost every time.

Once again, you must respond to the client's comments, and do it quickly!

Running Effective Client Meetings

Running an effective client meeting is a great way for a project manager to show he can successfully manage a project and guide the project towards the desired conclusion. On the other hand, I've witnessed many problems with client meetings, among them a lack of a purpose or even an agenda, discussions that go way off topic, poor time management, or post-meeting confusion about what happens next. To avoid these problems, an effective meeting process is needed.

The first step to have a successful meeting is preparation. Here are six "musts" for good meeting preparation:

1. Perhaps the most important consideration is to define the meeting objectives or expectations. I've been in many meetings where the objectives are assumed, and not stated clearly. This leads to a meeting that drifts, is not focused, and does not reach the goal.
2. Make sure the room is appropriate. Are there enough seats? Is a projector available, if needed? A flip chart? How does the lighting work? Don't hold meeting in a dark room to keep the slides visible.
3. Who should attend the meeting? Make sure they get invitations and homework assignments. Meetings accomplish more when the attendees are prepared. Tell them ahead of time what they need to read or do to be prepared.
4. Ensue that meeting attendees come prepared, including completing any homework.
5. Clearly state the date and time for the meeting and the length of the meeting. Start on time. Don't run over on the time. If attendees arrive late, talk to them after the meeting and tell them you expect a timely start.

6. What type of meeting is it? If the purpose of the meeting is to explore a topic, such as market trends in a particular industry, make sure everyone has an opportunity to contribute. Attendees should suspend judgment, since no decision will be made. If the meeting purpose is to make a decision, an agenda is a must, as is a strong leader or facilitator. Finally, if the meeting is being held to debate a topic, be ready for a competitive atmosphere, where learning is undermined. Determine ahead of time what type of meeting is needed, and prepare for it accordingly.

Next, a meeting should have a clear agenda. Here are a few pointers:

- What are your meeting objectives? Understand meeting purpose; use words like "plan", "identify", "develop", "recommend".
- What is the desired result? Be clear on what the meeting must achieve.... a list... a decision.... a specific action plan...etc.
- Get agenda approval by all attendees before proceeding. Revise the agenda if needed to get buy in.
- Discuss process & ground rules. Will you use a facilitator, recorder, timekeeper, etc.? Will leader act as facilitator? Hand out ground rules and let attendees review for a few moments.
- Conduct the work of the meeting. Use a facilitator, recorder and timekeeper, if needed.
- After the meeting work is complete, review action items and decisions. Issue assignments for the next meeting.
- Schedule the next meeting.
- Conduct a meeting critique. What can be improved? This is an important step if you want to steadily improve your meeting effectiveness.

Finally, the meeting leader or facilitator has several key responsibilities:

- **<u>Maintain a productive climate:</u>** It is the responsibility of the facilitator or leader to promote participation by all members, ensure that issues are understood by the group, encourage creativity and learning, and remove any barriers to progress.

- **<u>Maintain focus on meeting objectives:</u>** It is important that the facilitator or leader keep the group on the objectives. If potentially important sidebar discussions occur, put the topic on a parking lot for future discussion.
- **<u>Protect members from attack:</u>** The facilitator or leader must ensure that there is a climate where any ideas or proposals can be stated by the group without undue criticism or argument or personal attack, which will shut down participation.
- **<u>Direct processes that mobilize the group:</u>** The facilitator should utilize meeting processes that encourage participation and creative thought. For example, use a brainstorming process to get ideas and work towards solutions.

If you prepare for meetings, have a clear agenda, and run the meetings effectively, you will achieve the meeting goals most of the time!

The Project Management Framework

In this section, we present a Project Management Framework, which includes the key project management tasks every project manager should follow. But, before we get into the Project Management Framework, let's look at project management at the 50,000 ft. level for a moment.

What can go wrong when managing a project? Over the years, I've made my share of mistakes, and supervised many others. Here is my list of the most frequent project management problems, and what to do about them:

- Project management, from the perspective of the client, is all about *communication.* On time communication. Frequent communication. Frank communication. Clear communication. Communication that is too infrequent creates uncertainty and concern.
- A key responsibility of a project manager is to continuously *assess project risks.* Too many project managers totally ignore this subject, to their peril. When you don't look at risk, surprises occur. Almost all surprises that occur on a project are bad!
- Another problem that crops up a lot with project managers is attempting to *make decisions without gathering all of the facts.* Be careful

about rushing into a decision before assessing the facts. It is usually better to delay than make a poor decision.

- Project managers must *notify their clients of developing problems* at soon as they are recognized. Unfortunately, too many project managers first try to eliminate the problem, so they don't have to tell the client about it. However, in many cases, the problem just gets bigger. It is far better to inform your client even of small problem issues, and then demonstrate to them how you will solve the problems.
- What does a successful project look like? It is important for a project manager to *develop project metrics*, in concert with the client, that will show that the project is being performed successfully. Showing clients that key milestones/metrics are being met increases confidence and lowers uncertainty. Metrics could include milestone dates being met, budgets met, scope changes less than x, and so on.
- Finally, many problems have occurred because the *project manager did not have a good understanding of the work scope*. I use the word "understanding" here. It is the project manager's responsibility to be sure he or she has a clear understanding of every aspect of the work scope. The work scope must be consistent with the contract and consistent with the client's interpretation of the contract conditions.

Here's our Project Management Framework, the "magnificent seven" tasks:

1. **Project Management Plan:** A project management plan must be created at the start of every project. The plan should include metrics that define success, and creates a link between the contract and the work. The plan should include the work scope, schedule, budget, client contact information, client communication protocols, and project team. Each member of the project team gets a copy.
2. **Work Plan Development:** A detailed work plan is required to show a path to the expected results. A "meeting of the minds" with the client and the project team is necessary before beginning

work. The work plan is derived from the scope of work in the contract, and lists the work activities.

3. **Resource Procurement:** The project manager must be able to obtain and manage the human resources, physical resources, and other resources needed to carry out the project successfully. You *must* negotiate a contract with each team member. This involves scope, budget, schedule, and how budget overruns will be handled. When you do this, you significantly reduce the potential for project problems, and teamwork is enhanced because everyone understands their role. My experience is that negotiating budgets and schedules and scopes with *each team member* is one of the *most important* project management tasks there is!
4. **Project Budgeting:** The project manager must start with an *agreed upon* budget, then track expenditures throughout the project. Creating a work breakdown structure at the start of a project greatly simplifies budget tracking. Be careful of the Iron Triangle. If the budget changes, so will the schedule or scope. If the scope changes, so will the budget and/or schedule.
5. **Project Scheduling:** The project manager must create an *agreed upon* project schedule and monitor/update it frequently. Keep Peter Drucker's famous quote in mind: *"Time is the scarcest resource, and unless it is managed, nothing else can be managed."*
6. **Project Risk Management:** The project manager must alert the client about potential project risks and suggest mitigating actions early. We address project risk management in the next section.
7. **Project Quality Assurance:** The project manager is responsible for producing deliverables of the highest quality that "delight" the client by exceeding expectations (it is not only about "quality"). All projects should have adequate peer review. Larger projects should have a written Project Quality Plan to ensure proper focus on quality.

"Project Managers are like cats in a litter box. They instinctively shuffle things around to conceal what they've done." Scott Adams

Project Risk Management

Project risks are sources of problems for consultants. Every project has risks. Figure 4-2 shows ten sources of business risk.

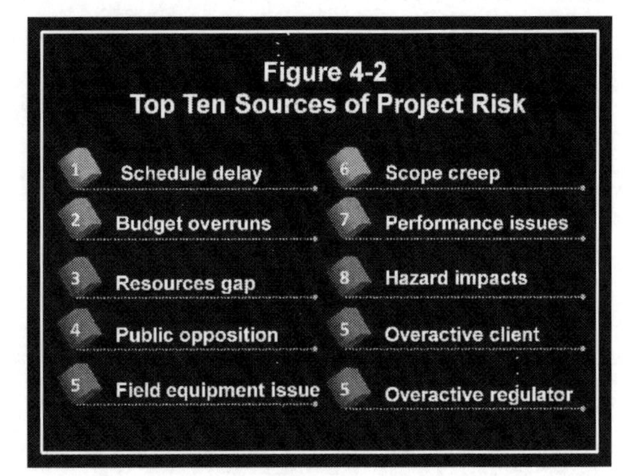

Project risk is defined as the exposure to unanticipated events that could negatively impact a project. Usually that risk is reflected in either increased cost or longer development timeframes, or both. More specifically:

Business Risk = f(likelihood of the event x the magnitude of the consequences)

An event may pose a high risk because of a high occurrence probability and high potential consequence, but we must be especially careful of risk events with a very low probability but very high consequences. An example is grain elevator dust explosions. If grain dust in a storage elevator reaches about 50 grams/cubic meter, with oxygen present in a contained space, and an ignition source occurs, a powerful explosion can take place that destroys the elevator and its contents. This is an example of a high consequence event with a low probability.

Consultants must be vigilant in identifying and monitoring risks, and mitigating them when needed. See Figure 4-3!

Project risk management is defined as the process of identifying key risk issues, evaluating the risk potential, and deciding if risk mitigation is needed. Getting back to the grain elevator, examples of risk mitigation are pneumatic dust control, and liquid (water or oil) additives.

Here is a three-step project risk management process:

1. Prepare a risk register at project kick-off. This is a simple task that can pay off big later on. It doesn't take much time to create this. Use it like a checklist. We'll show you an example of a risk register next.
2. Identify potential risks early. Looking for potential project risks must be on your weekly task list.
3. Discuss mitigation actions with the client. Always inform the client of any potential risks. When you do this, always suggest a risk reduction measure or options to reduce risk.

A risk register can look like this:

Risk Issue:	Biological resources found at the project site.
Consequence:	Possible mitigation of risks.
Initiating Event:	Raptor nests are found at the project site.
Liklihood:	25%
Risk Cost:	$500K-$1500K
Risk Mitigation:	1. Renegotiating land purchase for compensation 2. Abandon project site / find another project site

The definitions of the risk register components are presented below:

- **Consequences:** In the example, the potential consequence is "Possible mitigation" (this means the client may need to purchase additional land to offset impacts to biological resources, a potentially costly step).
- **Initiating Event:** What would cause the risk to occur? In the example, "Raptor nests found at project site" (many raptors are protected species).
- **Likelihood:** What is the probability that the initiating event will occur? In the example, we estimate a 25% likelihood.

- **Risk Cost:** What would the consequence be in cost to the client if the initiating event does occur? In the example, we estimate a cost of $500,000 to $1, 500,000 to purchase more land for mitigation.
- **Risk Mitigation:** How can you mitigate this risk to avoid some or all of the cost consequence? In the example, we could either renegotiate a land purchase to provide funds for land mitigation/purchase, or abandon the project site and turn to a back-up site.

Now, let's take a look at a few risk management techniques that can be applied at various levels of consequences.

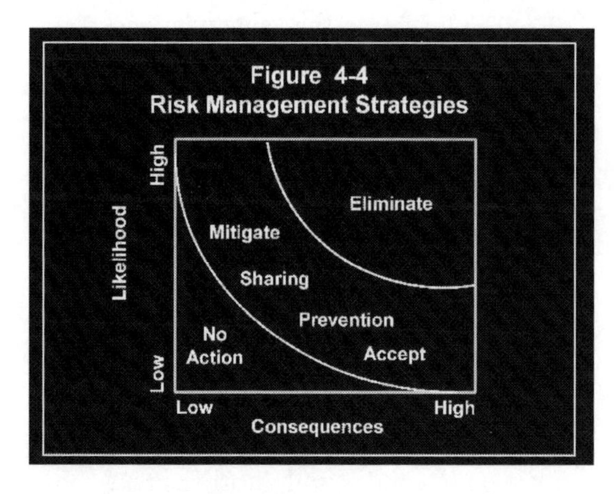

Figure 4-4 shows the risk space based upon the likelihood of an event (Y axis) and the magnitude of the consequences (X axis).

There are three basic responses to project risk: take no action because the risk is not significant; take some form of action for the risk; or eliminate the risk. Let's examine each response.

Risks that are low in consequences and low in likelihood will not require a response action.

Risks with moderate levels of consequences and likelihoods require some kind of response, such as mitigation, sharing, prevention, and accepting the risk:

- **Mitigation:** Mitigate the risk by implementing an action that reduces the risk to an acceptable level. Examples could include by monitoring, adding procedures, making process changes, using protective equipment, and writing contingency plans.

- **Sharing:** Examples of sharing the risk are purchasing insurance and public education.
- **Prevention:** Prevention is taking an action that prevents the risk from occurring. Examples could include using automation to replace a task, relocate a building that sits on contaminated soils, selling the asset with the risk, creating a new design that eliminates a risk, or improving a process.
- **Accepting:** You can accept the risk and its consequences.

Finally, risks with high consequences and/or high likelihoods require serious responses, such as eliminating the alternative with high risk (e.g. discard the alternative action with high risks). Risks can be eliminated by using automation to replace a task, relocate a building that sits on contaminated soils, selling the asset with the risk, creating a new design that eliminates a risk, or improving a process.

The challenge for the risk manager is to identify the events that trigger the risk, quantify the nature of the risk, and assess the consequences to the firm.

Service Innovation

We begin with a quote from Peter Drucker:

> *"There are only two functions of a business: innovation and marketing."*

This quote had a huge impact on me early in my career, and led me to focus on innovation and marketing as pathways to success in consulting. As a young consultant, I was fortunate to work with Dr. Ralph Keeney, who wrote the book on site selection for industrial facilities *Siting Energy Facilities* (Keeney, 1980). That experience gave me a great start in siting power facilities of all types. Gradually, I increased my expertise in siting, and coupled that with marketing to become quite successful in this field for decades. What made the difference was innovating a service over many years (i.e. continuous innovation).

There are many advantages to innovating services. When you offer an innovative service, you likely will differentiate yourself from your competition. You set yourself apart. You stand out.

When you are innovative, you are viewed as an expert. That is a good thing. Innovating tends to keep you current with new developments in your field. When you are thinking about innovative solutions, you become more focused on the needs of your client.

So, how do we innovate? Simply stated, we can innovate "inside the box" or "outside the box". Thinking inside the box means applying a systematic process, or traditional approach, to solve a problem. Also, thinking inside the box infers that you focus on the resources or tasks or processes that are close at hand.
Solve problems with the information and data and resources available, within the system in question. When thinking inside the box, it is very helpful to inventory the local resources you have, and the tasks and processes that reside in the system in question. Thinking inside the box does not depend upon the lightning bolt idea occurring.

Thinking outside the box means thinking with no constraints or self-imposed limits. It means thinking in non-traditional ways, such as having a mindset of openness to think differently. Thinking outside the box is realizing that new ideas may originate from outside the system under evaluation and even outside of the field of effort. Outside of the box thinking includes avoiding the use of formal problem-solving approaches or logic. Ask questions to create a loose boundary. In other words, use questions to provide some focus, rather than starting in a "blue sky" fashion. Limits help you organize thoughts.

Everyone has problems to solve! Harvard Medical School has performed research into how the brain solves problems. Let's say you are trying to solve a problem. Write down the problem just before going to bed. When you get into bed, think about the problem. Visualize the problem by using images as much as possible. Tell yourself to dream about the problem. Then picture yourself dreaming about the problem and compiling possible solutions.

When you dream, you enter a Rapid Eye Movement (REM) sleep. During this time, your subconscious, which never sleeps, will help create associations involving the images you pictured. REM sleep becomes a learning aid. For example, suppose you are having trouble focusing on a tennis ball as it hits your racquet. Imagine the ball, in slow motion, coming in contact with the ball and sinking into the strings, then being propelled outward toward your target. Imagine a specific target, such as a corner of the court. Do this 5-10 times before falling to sleep. Do this a few days and you will create a habit where you will find yourself following the ball into the racquet! This sounds strange, but it works. More importantly, if you are looking for a solution to a client's problem, define it as specifically as you can, using images as much as possible, and then tell yourself you need solutions. Essentially, you are programming your brain to help you. If you don't program your brain, you will not use the most powerful part of your brain, your subconscious. You cannot control the time it takes, but most of the time, you will get ideas.

Tina Seelig has performed research in creativity and innovation at Stanford University, and in one of her books, called *Insight Out: Get Your Ideas Out of Your Head and Into the World* (Seelig, 2015), she offers a framework for creativity and innovation:

Imagination = Creativity + Innovation + Application

Seelig says imagination is envisioning things that don't exist. Imagination expands by making observations, adding knowledge, and reframing problems. Think Polaroid for a reframe example. Think the Medici effect for the impact of knowledge on imagination. We apply imagination to become creative when we address a problem. Next, we apply creativity to generate solutions to problems. Finally, we apply the innovations to our work and bring them into reality. Seelig's book, *InGenius – A Crash Course on Creativity* (Seelig, 2012) is patterned after her course on creativity at Stanford.

Reframing is changing your point of view by asking "why". Asking why can expand the number of possible solutions to a problem as well as increasing the attractiveness of the solutions.

Here's a simple example of reframing. A business owner in Vermont asks a contractor to install an underground tank. Why? To store heating oil. OK, a more environmentally friendly way to store heating oil is to use an aboveground tank (a better solution). Next, why heating oil? Alternatively, electricity, propane, or cord wood could be used. A wood pellet stove is even more efficient and less costly. Once again, asking why reframes the problem and leads to better solutions.

Try reframing to create stronger value propositions, develop new services, or improve existing services.

The bottom line is that service innovation is a critical task for all consultants. Schedule time for creative thought. Don't do this with the television on, or with music playing, or with the radio on. Creativity tends to occur between your thoughts, or during quiet periods, when your mind is calm.

Before we leave the topic of innovation, I would like to discuss a brainstorming process that has been tremendously helpful over the years. It is based upon Dr. W. Edwards Deming's "Deming Cycle", consisting of four steps used to innovate: Plan-Do-Study-Act.

I call the process The Cool Idea Tool, or the P.O.S.E. Method: Problem, Options, Solutions, Execute.

It is best to start with a brief ice-breaking process, where the attendees might describe their expertise, their interests, and what they think they can contribute. The process leader should provide useful background information.

The first step is P, The Problem. State the problem context and its causes, in as much detail as possible or is known. State the facts, causes, and symptoms.

Get everything out. Write what you get on a flip charts. Hang them on a wall for all to see.

Next, go to O, The Options. This is the exploration phase. Now ask the group to offer ideas for alternatives, or solution ideas. Any ideas are OK, even apparently *outrageous* ones. No personal attacks are allowed, like "now that is a stupid idea", or "it will never work". Once again, write down all of the ideas on a flip chart and put it on the wall for all to see. It is important to build on people's initial ideas with wide participation by the group.

An option here is to conduct the ideation process in two steps: individually (attendees are in their offices), then in the group. This avoids possible inhibiting effects in a group, and possible groupthink fixation.

Next, go to S, The Solution. Either go around the room and require each person to suggest a solution (stressful on the shy), or just let anyone present (which usually means poor balance of participation). Either way is OK, although I prefer to put people on the spot rather than letting them sleep. Record every idea or solution in the speaker's words on a flipchart.

Now rank the ideas using the n/3 voting procedure (give attendees n/3 votes, where n= number of attendees). The solution with the highest number of votes is ranked first, and so on for all of the solution alternatives.

If there are a lot of solutions, try taking the highest ranked subset of the results and going through the process once more. So, if you had ten solutions in the first round, take the top five for the second round and repeat the process.

Finally, go to E, Execute. Perform the execution step, where you select the best solution, and develop an implementation plan. Is the final solution feasible? Is it operational? What resources are needed for implementation? Who will lead the effort? What is the schedule and budget?

This is a simple, yet very effective, idea generating, or brainstorming or decision-making tool. It is an excellent tool for use in groups. This remarkably simple procedure is a great way to establish a *consensus solution* in a group setting.

V. Key Skill No. 4: Problem Solving and Decision- Making

Key skill number four for consultants is problem solving and decision-making. Few consultants have received formal training in these subjects, yet problem solving and making decisions are skills that every consultant needs.

Introduction to Problem Solving

What is problem solving? One way to answer this question is to say that problem solving is the process of moving toward a goal when the path to the goal is uncertain. Problem solving is a *process*. Processes have inputs, outputs, and an aim or objective that must be attained by the solution.

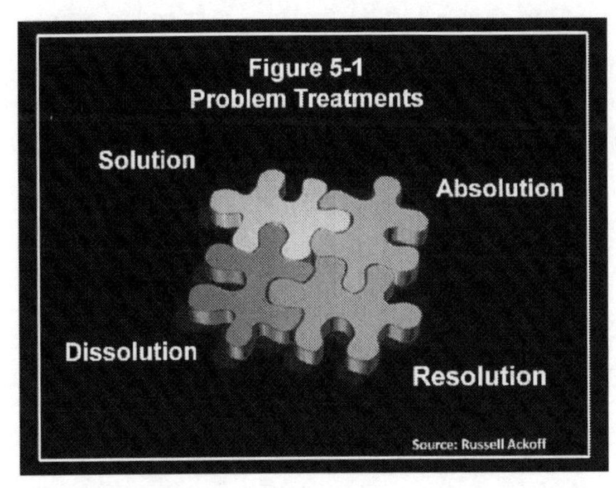

Russell Ackoff, in his book *Ackoff's Best: His Classic Writings on Management* (Ackoff, 1999) said that there are four high level ways to approach problems (shown in Figure 5-1): Solution, Absolution, Dissolution, and Resolution.

Each of the four approaches to problem solving is briefly discussed below:

- **Absolve.** First, we could decide to absolve the problem. That is, ignore it and hope that it disappears. This approach is similar to a sign in my dentist's office: "Ignore your teeth and they will go away!" We will not consider this approach any further!
- **Resolve**. Second, we could resolve the problem, which means that when we find an acceptable solution, we take it; however, the solution may not be optimal. Another definition of "resolving" is "satisficing". Satisficing seeks a consensus solution, *which can be useful in a group setting*. Satisficing typically limits the number of options…and suffers from time constraints…and typically limits data collection efforts. Satisficing can be viewed as a pragmatic, or

practical approach, especially when other methods don't seem to be working. Satisficing often turns into a rush to judgment, a quick answer. In summary, we want to avoid making decisions based upon "satisficing" because it is usually difficult to justify these decisions, and such decisions are likely suboptimal.

- **Dissolve**. Problem dissolution is a technique where you eliminate the problem by changing the system the problem resides in. A good example is Polaroid back in the 1970's. Before Polaroid, every camera used standard film you buy in rolls of 20 or more. You sent the used film to a developer for printing. Polaroid comes along and sells a camera that does its own developing and printing. This is a new system, eliminating problems associated with developing and printing, such as the fact that it could take several days to get your photos back from the developer. Problem dissolution involves a system re-design. Another example is what we could call "back room activities" like accounts receivable, invoicing, and payroll. Rather than dealing with these issues in house, you can outsource them to shops that specialize in these activities, and, therefore, they can conduct them at a lower cost. This solution is outside of the current in-house system. Thus, a common solution for a client is to outsource work to consultants, who may specialize in the task to be accomplished. In summary, when considering a problem, always look at the system, and see if the system can be changed to eliminate the problem.
- **Solve.** Problem solution is an approach where you optimize the problem solution. Rather than settling for a solution, we seek the very best solution, short of completely re-designing the system (problem dissolution). To optimize a solution, this usually means starting with what really matters to the stakeholders. For example, what is their attitude towards risk? What are their priorities? Are there any macro issues that will affect the problem solution, such as health and safety concerns, environmental policies or impacts, company mission, or economic cost issues? In other words, what is valued by the stakeholders? These issues can be lumped into what we can call "project objectives". In addition, we need an objective and logical decision process to determine which solution is best for the stakeholders (i.e. clients). We discuss this problem treatment further in this chapter.

Sources of Problem Solving and Decision-Making Biases

We are cognitive misers. In other words, we over-use System 1 thinking. We don't like to spend the extra energy that System 2 thinking requires. What do we mean by System 1 and System 2 thinking? Daniel Kahneman defines them in his book *Thinking Fast and Slow* (Kahneman, 2011). System 1 operates automatically and rapidly, with little or no effort and no sense of voluntary control. System 2 allocates attention to the effortful mental activities that demand it, such as complex computations or completing a tax form. We tend to use System 1 whenever we can because System 2 requires more resources (energy). When we use System 1, biases can creep into our thinking because our brain wants to make a mental shortcut whenever possible.

Here are eight common biases that you should monitor when solving a problem or making a decision:

- **Overconfidence in our solution.** We can convince ourselves that our solution is better than alternatives available. When we do this, we tend to limit the number of alternative solutions.
- **Groupthink.** A team may adopt groupthink by supporting each other regarding solution selection.
- **Risk Attitude.** We are typically risk averse. Our brains do not consider risk and uncertainty very well because that requires System 2 thinking.
- **Change.** Most of us dislike change, and will attempt to preserve the status quo.
- **Resource Consumption.** We tend to make decisions that conserve energy (System 1), especially if we are tired or stressed.
- **Ease of Evaluation.** We tend to avoid complex decisions in favor of straightforward decisions.
- **Poor Problem Framing.** We tend to adopt mental frameworks that simplify and structure the information facing us. This can lead to a biased view of problem understanding.
- **Anchoring.** The first data we observe tends to get higher weight in our minds. Also, past events or past solutions may influence our decisions.

In summary, review these biases often. Try to be as open-minded as possible when solving problems.

The Power of Questions

"The important and difficult job is never to find the right answers, it is to find the right questions." Peter Drucker said this to emphasize the importance of asking questions to solve problems. Asking questions is an important task to understand a problem situation.

Questions should address the problem objectives, methods and procedures, technology, communication, time, cost, people, and materials and resources. Each of these is briefly discussed below.

- **Problem Objectives.** Why is this task or process performed? What is the purpose of this task? What is to be accomplished? What are the required outputs? What is the result expected? What are the metrics describing the outputs? How will we know the result is attained?
- **Methods and Procedures.** Are the best methods being applied? Do we know if the methods *are* the best available? Can the procedures be simplified or altered to be more effective? Are the procedures being applied consistently?
- **Technology/Materials/Resources.** Is the appropriate technology/equipment being used? Is the equipment properly maintained or calibrated? Is the staff properly trained to use the technology? Are the required resources/materials available?
- **People.** Is the right staff assigned? Do they have the required expertise and experience? Is the workload assigned efficiently? Are the correct people working on the project or problem? Are there adequate people resources? Do we have sufficient labor hours and physical resources and equipment to solve the problem? Is management supportive & effective? Is there any data that contradicts or is inconsistent with current thinking?
- **Time.** Is the task within schedule? How can the task be conducted faster? Is the correct amount of time expended on the most important tasks? Are the tasks being conducted in the best sequence?

- **Cost.** Is the task being conducted within budget? Is the cost-benefit ratio attractive? Can the cost be reduced? How does the cost compare to the value-added?

Remember that questions are perfect for defining a problem and framing the problem. And, remember Rudyard Kipling's poem "Six Honest Men":

I keep six honest serving men
(They taught me all I knew);
Their names are What and Why and
When And How and Where and Who.

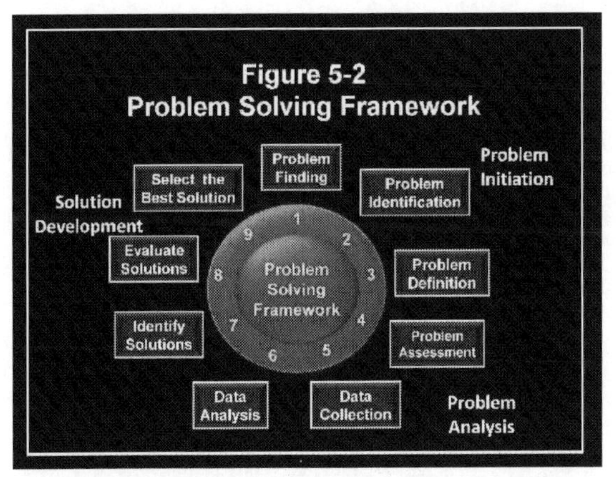

Now we'll introduce the Consultant's Problem-Solving Framework. I've used this framework for years and taught this framework to my staff when I was working for large consulting firms. This framework originated from the use of decision analysis in problem solving and decision-making. It is a very systematic and logical approach to problem solving. The framework is shown in Figure 5-2.

The first three steps are called Project Initiation, and include Problem Finding, Problem Identification, and Problem Definition.

The next three steps are called Problem Analysis, and include Problem Assessment, Data Collection, and Data Analysis.

The last three steps are called Solution Development, and include Solution Identification, Solution Evaluation, and Solution Selection.

Each of the nine steps in this framework is discussed below.

Problem Finding

The first step we call Problem Finding. Normally, most consultants don't think about this step. It happens automatically. However, putting your brain on auto doesn't necessarily lead you to the best solution!

Note that problem finding addresses both finding problems as well as *finding opportunities.* As consultants, we need to find opportunities! That is the *proactive* part of problem finding. The *reactive* part is largely driven by problems described by our clients.

The essence of *problem finding* is keeping an eye on the potential sources of opportunities or problems. If you are an investor in stocks, for example, you are devoting time researching new investment ideas and watching the economic trends. If you are a night watchman at a factory, you are constantly looking for problems. So, the key to problem finding is to be looking in the right places. What are your criteria for finding opportunities? What are you doing to find opportunities?

We can define two types of problems: the First Kind and the Second Kind.

- **Problems of the First Kind:** Some problems "hit you in the face, and you must address them immediately. No client wants to be put in this position. However, as we know, stuff happens. A crisis occurs, and the client must seek a solution quickly. Problems of the First Kind are the typical problems consultants see.
- **Problems of the Second Kind:** Some problems can be found because your scanning identifies them before they 'hit you in the face'. Note that problem opportunities are viewed as *more valuable by clients* because they *prevent* problems or solve them before they get *out of control*, or a system *is re-designed*, thus eliminating the source of problems.

How do we find problem opportunities, or in other words, how do we *find* problems? Here are four ways:

1. Find a gap in the project definition. Ask the client to describe the problem in detail. Ask "why" a few times. There may be un-met needs inside the project definition. The client may not have a complete understanding of their problem. I see this situation in

my site selection practice. Some clients like to locate a new power plant at an existing power plant site because the infrastructure is there. However, in some cases, these plants were built 30 years ago, and a town has grown around the plant. Adding a new unit will have significant impacts on the local residents, and therefore, could create a public relations problem.

2. Have all the project objectives been addressed? Most problems have more than one objective or goal. Often, we get anchored on one of the objectives. For example, when I become aware of a site selection opportunity, I ask the client "What are your objectives for this project?" I ask them to tell me how they selected the technology. I ask how they decided on the size of the project in megawatts. I ask the client what region of their service territory would be preferred.
3. Can a process be improved? In other words, perhaps altering the system will eliminate the problem and create an improved system as well. I've helped a few clients estimate the environmental risks at their industrial plants. Typically, these clients apply a very simple method to estimate risks that usually underestimates the risk costs, which can expose the client to unnecessary risks. I show them how to improve the way they estimate risks. Compare activities with established practices or best practices. Uncover resource deficiencies.
4. Prevention. Is there a way to change the process to prevent another problem from forming? One way to get these opportunities is to conduct audits. Use the findings to develop a proposal to solve potential problems.

In summary, your job as a consultant is not only to solve problems brought to you by clients, but to look ahead and anticipate situations facing clients that could develop into a risk or problem. An excellent discussion of problem finding can be found in *Decision Sciences* by Paul Kleindorfer, Howard Kunreuther, and Paul Schoemaker (Kleindorfer, Kunreuther & Schoemaker, 1993).

Problem Identification

Next up is Step 2 in the Problem-Solving Framework, called Problem Identification. Problem identification is defined as:

> *"A problem is identified when the gap between the current reality and the expectation reaches an action, or trigger, level."*

In other words, we can define a problem as a *gap* between the *current* condition and the *desired* condition. When such a gap is identified, a problem is identified, or found. Therefore, when evaluating a problem situation, or trying to determine *if* there is a problem, use the gap test.

Alternatively, a problem is identified, or recognized, when your scanning efforts show that a goal, need, value, criterion, objective, or social norm is outside of its acceptable limits. For example, several sites are being considered for a new manufacturing facility. One selection criterion is to have a major university located within 25 miles of the facility. Site 6 is located 40 miles away. Therefore, the site must be dropped from consideration because it is outside of acceptable limits.

In summary, problem identification requires some type of triggering event that identifies a situation or condition, as a problem.

Problem Definition

Step 3 of the Problem-Solving Framework is Problem Definition. In this step, we must develop an accurate and complete problem statement or understanding. Creating a strong definition is a critically important step. Don't simply accept a client's definition. Go deeper than the client. Ask questions. Probe. Recall what Einstein said about a problem definition. He said if he had only one hour to save the world, he would spend 55 minutes on the problem definition and 5 minutes on the solution.

Typically, we want to address the problem domains: the background, system, the goals or objectives or triggers, and the environment.

- **The Background.** Describe the situation leading up to the problem. What events took place that appeared to contribute to the problem? Clarify the difference between the problem and symptoms of the problem.
- **The System.**_Describe the system the problem resides in. What is the aim of this system? Describe the higher-level system. What are the individual processes within the system? How do you measure the system output? What are the constraints on the system? What people are in the system? What controls are present, such as procedures? Note that examining the system is another way of saying to examine the frame of the problem.
- **The Goals.**_What are the goals or objectives? How are the objectives being compromised by this problem? What is the gap between the current situation and the desired or ideal situation?
- **The Environment.** Describe the environment surrounding the problem and system. What external factors could impact the problem, such as weather, other people, the business environment, or other possible constraints.

Here is a series of questions that can aid development of a problem statement:

- **What are affected by the problem?** For example, a slower economy creates higher unemployment.
- **How is it affected?** High unemployment creates problems in the economy.
- **Who is affected by the problem?** Primarily the middle class and the poor.
- **When is it a problem?** When unemployment reaches 8%.
- **Where is it a problem?** Across America.
- **Is the problem getting worse?** Yes.
- **What assumptions surround the problem?** Challenge these assumptions.

- **Does a slower economy always lead to higher unemployment?**

Most problem definitions should have three components: a subject, a target metric, and a direction. Here's an example:

Our problem, the subject, was that vendor invoices were taking too much time, i.e. it often took more than 30 days to get vendors paid. This is the trigger.

This created two problems for us: first, some vendors stopped accepting orders from us, which created more work for us to find another vendor. Second, we were outside of our contractual conditions, not a place we wanted to be.

The target metric is that we must pay vendors in less than 30 days.

The direction is to *reduce* the time vendor invoices get processed.

These three steps complete the first part of the Problem-Solving Framework, called Project Initiation. These are preparatory steps that must be performed carefully and in detail. *Without this preparation, it is likely that the solution to the problem will not be optimal.* Taking the lead from Einstein, 90% of the problem-solving effort is in the preparation.

Problem Assessment

The next three steps in the Problem-Solving Framework are called Project Analysis, where the problem is evaluated. The steps are Problem Assessment, Data Collection, and Data Analysis.

In Problem Assessment, we look at the symptoms, causes, and root causes that are related to the problem. Essentially, we want to understand the relationship between the causes and the problem, being careful to distinguish between causes and symptoms.

Figure 5-3 shows one way to look at problem assessment: use a systems view.

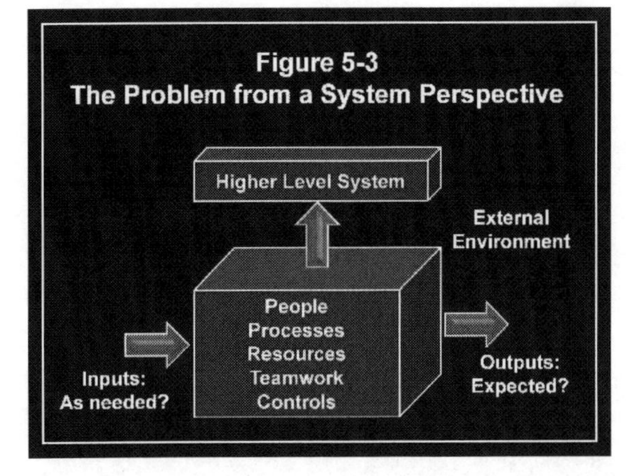

Take the perspective that the problem resides within a system.

Evaluate process inputs and outputs. Are all of the inputs as needed? Are other inputs required to solve the problem? Are the outputs as expected? Do the outputs meet performance goals?

Within the process where the problem resides, consider:

- **People.** Is the appropriate staff in place? Do they have the training they need to be successful? Are there any staff needs that aren't being met? Can the staff be reduced without impacting performance?
- **Processes.** Are the processes functioning properly? Are there any bottlenecks that affect performance? Are there any redundant processes?
- **Resources.** Are there sufficient resources to meet work demands? Are the proper technology resources in place to enhance productivity?
- **Teamwork.** Is the staff working together and communicating effectively? Do they draft team goals? Do they specify appropriate roles and responsibilities? Does the team handle conflict effectively?
- **Controls.** Are there procedures that support success? Is management supportive? Are there any other constraints that inhibit attainment of performance goals?

Problem Assessment usually includes a Root Cause Assessment. The issue here is to identify symptoms, causes, and root causes, *and be able distinguish between them.*

Let's suppose you are not getting enough clients in your consulting business. A *symptom* could be a lack of good potential leads in your sales funnel.

Another symptom could be a poor conversion rate. Clients are interested, but they just don't buy.

A *cause* could be that these prospects are not being convinced or sold properly.

A *root cause* could be poor targeting in your market.

A solution could be to prepare a marketing plan based upon solid market research and achieve a better focus on your target prospects.

Root causes are typically harder to diagnose because they are hidden deep within a system. Budget overruns, loss of clients, decreasing market share and high staff turnover are easy to see, but they are symptoms of the problem. The root causes could be poor assumptions, poor strategy, poor implementation of a strategy, poorly trained staff, unclear roles and responsibilities, unrealistic goals, or ineffective management.

Problem Assessment begins with asking the right questions (have you heard this before?). Here are some simple example questions that will help assess a problem:

- **What data will help you describe the problem**? This question relates to the type of data and information that could help assess the problem. Are there reports? Field data? Studies that have been published? Surveys completed?
- **Where can we find pertinent data?** It is always helpful to seek all the data relevant to a problem.
- **Where did the problem take place?** This may be relevant.
- **When did the problem occur?** This is very important. If the problem is new, it is unlikely that the client has evaluated the problem yet. However, if this problem has existed for a year, there is likely to be a lot of data available. Also, knowing when may lead to a cause.
- **Who is impacted by the problem?** This is always a good question. Who are the stakeholders who need to be included in the problem analysis?

- **Why is this problem important?** This question leads you to why the problem should be solved and who the stakeholders are.
- **How can we determine the cause?** As we mentioned earlier, most problem assessments include a cause and effect analysis to determine the underlying cause or causes.

Finally, there are a number of analysis tools that can be helpful for conducting a problem assessment.

Examples are simple brainstorming (such as the Cool Idea Tool we described previously), a Fishbone Diagram (also called a cause and effect diagram), a force field analysis, or the Five Why assessment (ask "why" at least five times to dig deep into the problem causes).

Data Collection

The fifth step in the Problem-Solving Framework is Data Collection. In many problem situations, you need to collect some data. For starters, here are some suggestions regarding data collection:

- Make a clear path between data and the problem. In other words, make sure that the data are correlated with the problem.
- Is the data time sensitive? This could mean two things. First, is the data still good, given its age, or is the data only applicable to certain times? Biological data, for example, is typically time sensitive, because it must be collected in the Spring.
- Be willing to challenge the data integrity. In other words, make sure that the data is correct.
- Know the source of the data. This is always a must to be sure your data are credible.
- Be aware that data comes in layers. For example, data can be "visible" and "invisible", or obvious and not obvious. Many datasets were constructed from an underlying dataset.
- Identify data inconsistencies and errors. Always proof the data.
- Is the data a contributor or symptom? Make sure that the data is related to the problem, and not to a symptom of the problem. Data describing the amount of water collected inside a basement

describes a symptom. Rather, you need structural information about the basement walls, and how rain water flows or collects outside of the basement.

- Categorize the data into "soft" (qualitative) and "hard" (quantitative).
- Organize the data with a timeline. This can be very helpful, especially for complex problems. For example, data timing is critical when assessing soil or groundwater contamination, which changes over time.

Here are six data collection methods that I've used extensively:

- **Observations.** Examples of observations include surveillance cameras in stores, or accumulating traffic counts at a street. It is usually easy to make observations. When evaluating potential sites for industrial development in the desert, I always try to be present during a significant rainfall, to observe how much water (depth and velocity) enters the project site, and how it flows across the site. In many cases, simple observations are often better than spending a lot of money modeling a situation.
- **Surveys.** Examples of field surveys include biological surveys, water quality surveys or testing, or farm management surveys. Often, surveys are required to obtain a conditional use permit or a construction permit.
- **Interviews.** Examples of interviews include phone interviews or job interviews. We use phone surveys frequently to gage public opinion regarding a proposed project. You can engage a firm to conduct a phone survey quickly and at low cost. Then you can share the results with electeds during the project's permitting process.
- **Document or Data Review.** Examples include data rooms used for sales of projects or acquisitions, or searching for and reading available data.
- **Questionnaires.** Examples of questionnaires include medical questionnaires, community surveys, client satisfaction questionnaires, or job applications. This is another technique that

is relatively easy to conduct, especially with software tools like Survey Monkey. Just be careful that your questions are not biased.

- **Public Meetings.** Holding public meetings is a great method for eliciting the public's attitudes regarding a proposed project, or impacts of the project or proposed locations for the project.

Data Analysis

The sixth step in the Problem-Solving Framework and the third of three steps in the Problem Analysis group is Data Analysis. We'll briefly describe seven common data analysis tools. We delve deeper into the analysis of uncertainty in data.

- **Simple Statistics.** Simple statistics include means, averages, and standard deviation. Simple statistics aid in understanding a data set.
- **Diagnosis Tools.** Examples include a Force Field, and Cause & Effect Diagrams.
- **Tabulations.** Check sheets and comparison matrices are examples of data tabulations.
- **Graphic Analyses.** Histograms and scatter-plots are common graphic analyses, as are control charts.
- **Multi-variate Statistics.** Analysis of variance and regression analysis are examples. Analysis of variance, or ANOVA, is a statistical test of whether or not the means of different data groups are equal (for example, the difference in age distributions). In regression analysis, a dependent variable is assumed to be a linear function of one or more independent variables. Regression is used for prediction and forecasting. For example, forecast auto sales based upon personal income. Is there a relationship between these variables? Or, is there a correlation between the value of the dollar and the price of commodities?
- **The Data Analysis add-in in Excel.** This is a great tool that will perform many of these analyses, such as simple statistics, ANOVA, Regression, histograms, and scatter plots.

- **Handling Uncertainty in Data.** In many cases, the data are uncertain, yet we must learn from the data. It is all we have available. The treatment of uncertainty in data is discussed below.

We cover two common ways to evaluate data uncertainty. By making the uncertainty *explicit*, we can estimate a range of expected outcomes. One way is to use a probability distribution rather than a single value to represent an input parameter. We discuss this approach later in this chapter (Simple Additive Decision Model). A second method to handle uncertainty is Monte Carlo Simulation. I've used this technique hundreds of times over my career in consulting, and found it to be one of the most useful analysis techniques for helping clients understand risk and uncertainty.

There are many applications of Monte Carlo simulation in engineering consulting and business consulting. Here are just a few examples:

- Remediation cost estimation
- Evaluation of remediation alternatives
- Environmental claims management
- SEC disclosure requirements for environmental liabilities
- Strategic planning/resource allocation
- Analysis of risks in mergers and acquisitions
- Estimation of major equipment upgrade costs
- An adjunct to Due Diligence, where a clear financial liability is expressed
- Estimation of reserves needed for environmental liabilities

Monte Carlo Simulation is used to approximate the probability of certain outcomes by running multiple trial runs, called simulations, using random variables, or random numbers, which are used to calculate a result see, Introduction *to Simulation and Risk Analysis,* (Evans & Olson, 1998). For example, a remediation cost is calculated by multiplying a probability times a cost. The simulation runs 1000 trials multiplying a probability (say, .75) times a cost as defined by a cost distribution, say a lognormal distribution with a 50% value of $100,000 and a 90% value of $225,000. These two values define the distribution. 1000 trials are made using this distribution. The output is a cost distribution. Next, we'll look at an example.

Figure 5-4 shows a Risk Management Framework, an eight-step process to conduct an assessment of project risks using a simulation. Each of the steps is discussed below.

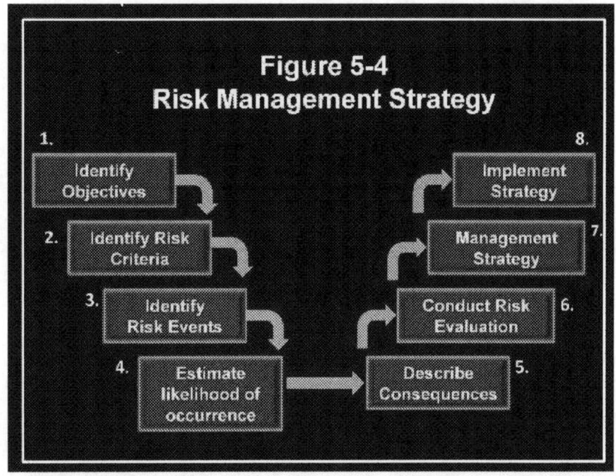

- **Step 1: Identifying Objectives.** Identify performance objectives, based upon the firm's business objectives, identify stakeholders and their interests. Examples are how much to reduce a bid price to account for acquired risk, estimating insurance claims for negotiation of insurance claims, or assessing the consequences (e.g. mortalities) from an event. The objectives must be specific and measurable.
- **Step 2: Identify Risk Criteria.** It desirable to have a set of criteria against which risks can be evaluated to determine risk acceptability. For an acquisition of an old industrial site, a criterion could be cost of subsurface remediation. For health risk, we could use a cancer risk probability, such as 1×10^{-4}.
- **Step 3: Identify Risk Events.** Events are identified from past experience, modeling, studies, and expert judgment. An example might be a transformer oil leak from an electrical substation. Or, or air emissions from a storage tank. Or, an upset condition at a manufacturing plant. For example, when evaluating risks at a plant, a "risk register" is compiled. The register includes a brief description of the risk issue, its potential consequence, the probability that the issue will occur, and an input cost distribution.
- **Step 4: Estimate the Likelihood of Occurrence of the Risk Events.** What is the probability that a risk event will occur? An example might be "there is a 5% probability that one hazardous waste transport truck in the fleet will have an accident on a year that results in a release." Or, "there is a 25% probability that an underground tank spill will require remediation". Risk event

probabilities are estimated by an expert panel with relevant experience.

- **Step 5: Describe the Consequences of the Risk Events.** What is the consequence of the risk event? A truck carrying ethylene crashes, causing a BLEVE (a Boiling Liquid Expanding Vapor Explosion). Consequences include mortalities and injuries to individuals within the blast radius. For an oil spill from a storage tank, characterize the nature of the required remediation.
- **Step 6: Risk Evaluation.** Quantify the risk by calculating expected values of risk -- probability times consequence -- and create a risk profile which shows each risk event ranked in order of decreasing risk, and an exposure profile, which shows the range of consequential cost for the ranked cost events. Costs for environmental issues are typically expressed using a lognormal distribution. Also, costs are spread over a 15-year time horizon, over which expenses can occur.
- **Step 7: Develop a Risk Management Strategy.** If a risk is detected, what is the strategy for managing this risk? For example, we can consider transferring the risk (for example, sell the division that has the risk), reducing or mitigating the risk, or just accepting the risk and its potential risk costs.
- **Step 8: Implement Risk Strategy.** In this last step, the risk management strategy is implemented, and the results are evaluated to determine if the strategy accomplished its risk reduction goal.

We'll use an example of a liability assessment for soil and groundwater contamination at an industrial facility to demonstrate the Monte Carlo simulation. More specifically, we will consider a No. 6 oil spill at the facility.

First, following the Risk Management Framework, what is the objective of the liability assessment? In this example, the objective is "identify potentially significant residual liabilities", including the No. 6 oil spill.

Second, identify risk criteria associated with soil and groundwater contamination. In order to make decisions, we need to have a set of criteria to determine whether the risks are acceptable, or should be mitigated. Risk criteria could be a cost trigger, for example. Or, a risk trigger could be a risk level, such as a probability of 1 in a million, or 1 in 10,000. For this example, a report of an oil spill is the trigger event.

Third, describe risk events arising from existing contamination at operating plants. For example, what types of events could occur that would result in a release to the environment. Examples could be an oil leak from an electrical transformer or a spill from a gasoline storage tank. Once again, in our example, the risk event is the No. 6 oil spill.

Fourth, describe the likelihood of occurrence of the events. How do you estimate these likelihoods? Actuarial data is available for accidents. Estimates of flood frequencies and other dangerous weather-related events, for example, are available from FEMA and NOAA. Knowledgeable employees or consultants are often excellent sources of likelihood and cost data.

In our example, since an oil spill was identified, the probability of this event is 100%. Clearly, many events have lower probabilities. For example, another risk event for this facility was a tank leak which contaminated soil and, perhaps, groundwater. The likelihood of soil removal was 100% and the probability of product recovery from the groundwater was 60% based upon existing sampling data.

Fifth, the value, or cost, or magnitude, of the consequences of risk events must be estimated. Since we are attempting to estimate the cost of events that haven't occurred, the estimate will be uncertain. Therefore, we typically account for the uncertainty by expressing the cost in terms of a probability distribution. For environmental risks, the lognormal distribution is often used because it is skewed to the right, or high cost tail to account for the uncertainty. In our example, the estimated cost at the 50% probability level was $75,000 and the estimated cost of cleanup at the 90% level was $200,000. In other words, there is only a 10% chance that the cleanup cost for the oil spill will be $200,000 or greater. We use the 50% and 90% probabilities because that will define the shape of a lognormal distribution.

Sixth, the risk is evaluated by developing a risk profile using Monte Carlo simulation. A risk profile shows the relationship between risk events and how the total risk is distributed across all risk events. A typical risk profile ranks risk events in order of decreasing risk cost, thus showing which events are riskiest. We will show an example of a risk profile for our No. 6 oil spill shortly.

Seventh, the risk profile is used to create a risk management strategy. The purpose of a strategy is to reduce exposure to risk, or more generally, to elucidate the corporate or facility risk exposure. If the risk cost, as shown in the risk profile, is sufficiently elevated, a strategy is developed to mitigate the identified risk.

Here's a brief introduction to Monte Carlo Simulation. Simulation is useful when problems exhibit significant uncertainty. This approach is an excellent tool for estimating the cost consequences of risk events under uncertainty.

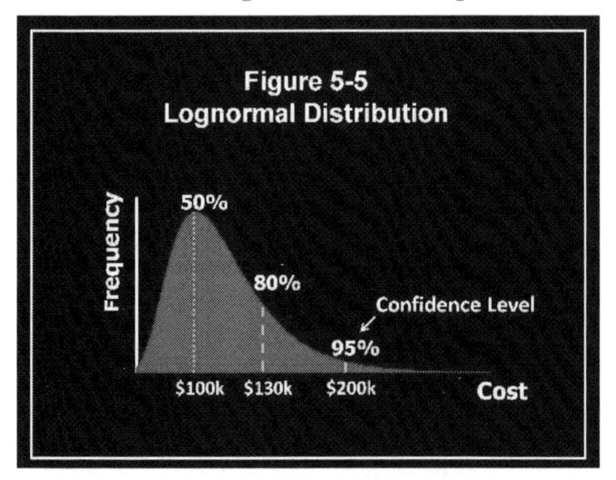

Monte Carlo simulation is basically a sampling experiment, run with about 1000 to 2000 trials, conducted to estimate the distribution of an outcome variable that depends upon one or more probabilistic input variables. Simulation, then, provides an estimate of the magnitude of consequences of risk events.

Typically, for environmental issues, we are interested in the 50%, 80% (some prefer to use 75%), and 95% values in the outcome distribution, as shown on Figure 5-5. The simulation output costs are assembled in a distribution, and we obtain the three costs we want from this distribution.

The 50%, 80% and 95% probability level costs are used to create a risk profile. An example risk profile is shown in Figure 5-6.

Figure 5-6 shows the complete risk profile for the industrial facility liability assessment we used as an example throughout this section. This is a plot of "Occur Cost" versus all of the risk issues identified at the facility. The occur cost is the estimated cost if the risk event occurs. The simulation uses a probability distribution to calculate the occur costs. Consider the leftmost bar in Figure 5-6. The top section of the bar indicates the 95% confidence level occur cost (about $60,000,000), the middle section of the bar indicates the 75% confidence level occur cost (about $29,000,000), and the lower section of the bar indicates the 50% confidence level (about $17,000,000). The highest bars signify the greatest risk costs. The line shows a relative comparison of risk exposure called a "risk quotient", calculated as a likelihood multiplied by a cost. This risk profile shows that the two leftmost issues contribute the majority of the risk.

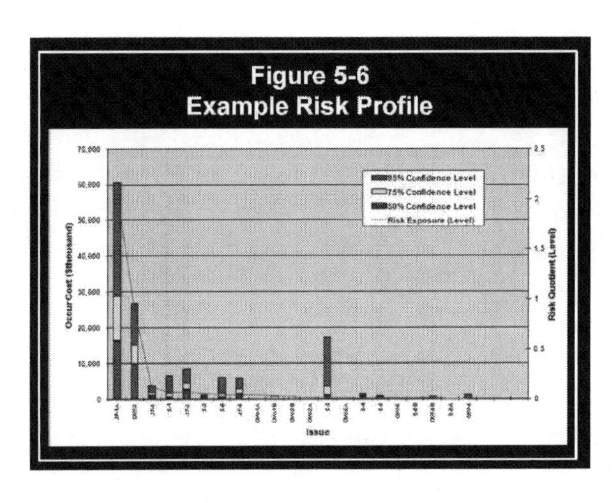

Figure 5-6
Example Risk Profile

A portion of this profile is enlarged in Figure 5-7.

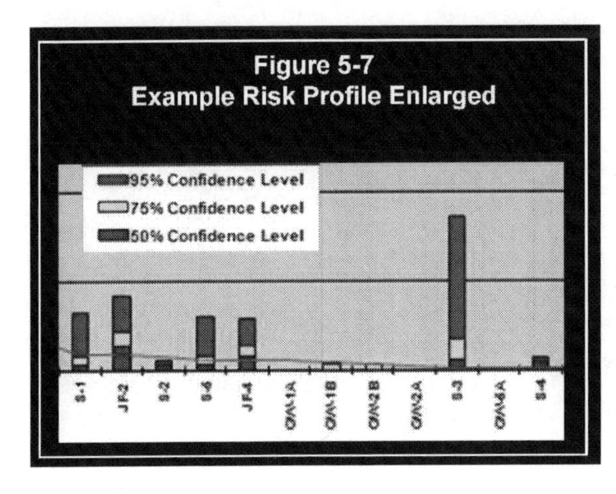

Figure 5-7
Example Risk Profile Enlarged

The enlarged view of a portion of the risk profile shows our example, No. 6 Oil Spill, designated as Issue S-1, on the left. The risk of this event is relatively small in comparison to some of the other events shown in Figure 5-6.

In some cases, we could add a horizontal line on Figure 5-6 indicating the level of acceptable risk (or below).

With the client armed with the risk profile, he can decide which risk events require immediate mitigation and which risk events can be mitigated over a longer time frame.

In summary, the Monte Carlo simulation is a very useful tool for dealing with uncertain data. There can be a lot of valuable information embedded within uncertain data. This is a tool that can be used to extract that information. Regarding software that runs these simulations, two of the most popular ones are Oracle Crystal Ball and @Risk. I was fortunate to work with Adrian Bowden, who applied these techniques to environmental risk problems. His book, *Triple Bottom Line Risk Management* (Bowden, Lane & Martin 2001) is an excellent read on this subject.

Identify Solution Alternatives

The seventh step in the Problem-Solving Framework, and the first step in the Solution Management Group is Identify Solution Alternatives. *This is an important step, in that the quality of the problem solution will be a function of the quality of solution alternatives identified.* In my experience, the lack of a significant effort to uncover potential solution alternatives is one of the typical reasons why problems are not solved in an optimal way, and the client is not well-served.

First, we say "make a broad reach for alternatives", because in many cases, the problem analyst decides prematurely on a solution. We can short circuit a solution alternatives task if certain biases enter the alternatives effort. For example, be careful about relying on options used in past projects. Just because a solution worked on an earlier problem doesn't mean it will work on all problems. Also, the first option identified can anchor or influence the subsequent ones. We humans are lazy. We often will settle for the first solution that comes to mind. Our brains don't like ambiguity and uncertainty.

Second, try reframing the problem. Does that expose new alternatives? The simple process of asking 'why' questions provides an amazingly useful tool for expanding your list of solutions for a problem. Another way of reframing is to list the assumptions and challenge them. We were asked to identify a new waste treatment facility for a town. We looked at one new technology that appeared attractive; however, it was expensive. Our assumption was that the residues would be landfilled, the usual solution. We challenged that assumption and determined that the residue would be fine for road base. That helped the economics significantly.

Third, and most importantly, think about values first, or what is truly important to you or your client. Ralph Keeney in his book *Value-Focused Thinking* (Keeney, 1992), mentions two ways to develop alternatives: Alternatives-focused and objectives based. Alternatives-focused means going directly to developing the alternative solutions once you are faced with a problem. Alternatives-focused thinking is *reactive,* in that you are confronted with a problem, and you must generate some solution options. Sources of alternatives-focused solutions include past experience and reaching for the obvious solution, as we mentioned earlier.

Objectives-focused thinking, on the other hand, is *proactive*, in that considering objectives, or values, early in the problem assessment process, you may find other problems, or may be in a position to better define the problem. By finding other problems, I mean applying a value system may lead to identifying improvements that will provide a better result than anticipated.

Objectives-based means starting the alternatives development process by considering the *project objectives*. What do the decision makers want out of the project? Assume you are working on a new building to house the town's symphony orchestra. What is an objective? Increase the quality of life for the citizens? Attract more investment on Main Street? A top objective is not to ensure that the building complies with seismic standards!

Or, consider a municipal waste incineration system. One objective may be to reduce byproducts as much as possible because they can be costly to dispose. However, this may prompt a look at productive use of the byproducts, which could increase revenues and make the entire project pencil out better and be a financial winner.

You will find that considering objectives *before* listing alternative solutions will broaden your list and *stimulate deeper, better thinking*. In fact, it's likely you will discover more effective solutions by focusing on objectives first.

Where do objectives come from? The primary source is the decision-maker's values, or put another way, what the stakeholders think are the most important issues regarding the problem at hand. These issues, for example, may be related to policies in government or regulations, or laws. Other sources could be past experience, output from brainstorming sessions, or from strategies or goals the firm has developed.

One tool that has been very useful for the alternatives search is an "Objectives Hierarchy". This tool helps add some structure to a list of project objectives. Which objectives are "top line" objectives or "end" objectives, and which ones are subordinate to others ("means objectives" or lower level objectives). The top level of an objectives hierarchy is the all-inclusive objective, and is broad in scope and indicates the real reason for the interest in this problem. Note that there is no "perfect" structure.

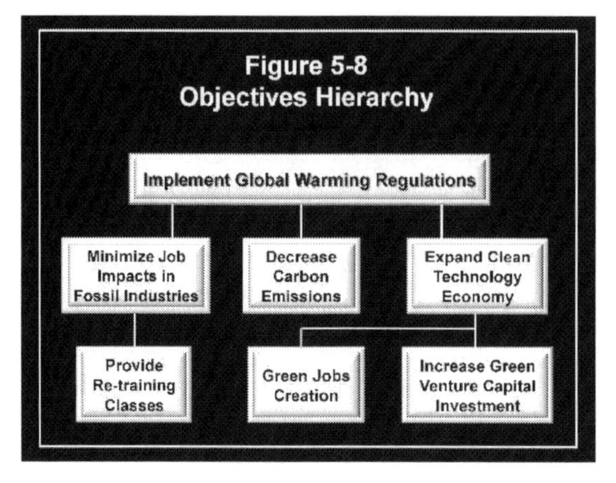

An example is shown in Figure 5-8. The top objective is "Implement Global Warming Regulations". This objective is broken down into three sub-objectives: "Minimize Job Impacts in Fossil Energy Industries", "Decrease Carbon Emission", and "Expand the Clean Technology Economy". How do you implement these sub-objectives? That brings you down to the next level. Minimize job impacts in the fossil industries *by providing re-training.* Expand the clean technology economy by creating more "green jobs" and increasing green venture capital investment.

In summary, creating a long list of alternatives is a critical task in problem solving. The best way to create such a list is to develop the project objectives first, and let them guide you towards the appropriate solution alternatives.

Evaluate Solutions and Select the Best Option

The last two steps in the Problem-Solving Framework are Evaluate Solutions and Select the Best Option. These steps are conducted together. Once a decision tool is used to evaluate the alternative solutions, the preferred solution should be evident. The last three sections of this chapter present a variety of decision tools that can be used to decide which solution alternative is preferred by the client.

How Decisions are Made

When people make decisions, a number of factors may impact, or influence, how these decisions are made. It is helpful to be cognizant of these factors when deciding. Here are a few factors to watch for:

- **Past Experience.** Most importantly, your past experience will shade your decision making. This is so because your brain accesses this information to help you decide. Fortunately, our brain is wired to not repeat the mistakes of the past. Remember that when working with clients, their decisions are colored by experience as well. For example, I was working on a reservoir project, where we needed to conduct outreach to the public about the project. At first, our client did not want to do this because in a previous project, the public killed his project. I had to explain that his project failed due to poor siting, not because he conducted public outreach.
- **Values**. Your values will influence how you decide, and your client's values will impact their decisions. In fact, we say that these values need to be out front when decisions are made. If values are not part of the decision, the quality of the decision will surely suffer.
- **Available Data.** Decisions will be dependent upon the data at hand. Therefore, always be sure that the data set is as complete as possible. Extract as much information from the data as possible.
- **Skills of the Team.** The skills of the decision-makers will govern, to some extent, how decisions will be made. This is especially important with respect to skills in decision-making. This is why we spend time on this subject. You, as a consultant, must lead your client in the decision-making process.
- **Systematic versus Intuitive.** There are two parts of our brain that help us to make decisions: the intuitive side, or the intelligent intuition side as described by Jonah Lehrer, author of How We Decide. Our rational brain focuses on the facts of the problem. People use these faculties in different ways. The

key point is that *every* decision is based, in part, on your feelings.

- **Detail versus Pattern Oriented.** As you know, some people like to get lost in the weeds of detail, where others prefer to stay aloft and just look at higher level patterns or rely on their intuition. This will impact how people make decisions!

Decisions are made based upon some type of decision rule. The simplest rule is intuition, human judgment, or "gut feel". These kinds of decisions are made based upon past experience. Typically, these decisions are a rush to judgment. Another rule is base the decision on the facts at hand. For example, if all of your decision criteria rank best for one alternative, that alternative "dominates", and the decision is a sound one. Or, you can "eliminate" alternatives that do not fall above a threshold. For example, if you want to purchase a new car, and your budget is $15,000, you will eliminate most cars due to cost. In a third approach, called "lexicographic", you decide which decision criterion is most important, and select the alternative that is best on that alternative. If there is a tie, go to the second most important alternative. These are examples of simple decision tools. Obviously, they do not apply to complex decisions.

For complex decisions, we turn to analytical tools and models. For example, if cost is the key decision criterion, the discounted cash flow method can be useful. Cost-benefit analysis is another tool when cost is an important consideration, as well as decision trees. In addition, there are many other decision tools that are available for situations where cost is not the main issue.

How do we decide which decision tools is best? Consider the following questions:

- Who are the decision-makers that must be satisfied? Are they more accustomed to intuitive or human judgment decision making, or are they comfortable with more formal methods? Do they expect that values should play a role in decisions?
- What is the desired outcome? Looking for an optimal solution? Is the goal to create a defendable solution that can withstand scrutiny? Is the decision-making process being

conducted to justify a prior decision? Do you need a precise decision? Is the problem complex and deserving of a more analytical approach (e.g. decision analysis), or can simpler methods be applied (e.g. comparison matrix)?

We will discuss two decision tools in the following sections of this chapter. The first is pairwise comparison, and the second is decision analysis. I have found these tools to be extremely helpful for consulting engineers.

Pairwise Comparison

Pairwise comparison is a technique I have put to use many times in my career. It is a quick and easy tool to use when you need to reduce the number of alternatives, or options from a large number (say 20) to a smaller number (say 5-10). Then you would use another tool such as the comparison matrix or decision analysis to get to a selected solution. Pairwise comparison is great for comparing many options, but I wouldn't use it to decide amongst just a few options (especially less than five). This method is great for getting a rough ordering of alternatives. Note that only one criterion is used. In addition, pairwise comparison is not a values-based method.

Pairwise comparison works like a round-robin tournament in which every candidate/alternative is matched one-on-one with every other candidate. The candidate with the majority of wins is the preferred candidate/alternative. The procedure is as follows:

- **Step 1:** Apply one-on-one comparisons amongst all alternatives.
- **Step 2:** Prepare a ranking table.
- **Step 3:** Make a choice between every matchup. Assign one point to winners, ½ point to each alternative for ties, and zero points for losers.
- **Step 4:** The best alternative is the one with the highest number of wins (points).

Note that pairwise comparison only provides an ordinal ordering, that is the intervals between each candidate/alternative are not necessarily equal.

Pairwise comparison was the decision tool of choice when I performed a landfill site selection study in the Midwest. About 25 potential landfill sites were identified in the first siting step. We needed a way to quickly eliminate the least attractive sites. Note that only one criterion is used to decide the comparisons. However, you can define the criterion such that more than one issue can be included. In the landfill study, we defined the criterion to eliminate sites that were close to populated areas, sites that were in or close to flood zones, sites that were adjacent to highways, and sites with unacceptable topography.

Let's look at a simple example. In this example, we have four job offers (recall that we don't recommend pairwise comparison for four alternatives). The applicant decides to use pairwise comparison as a decision-making aid.

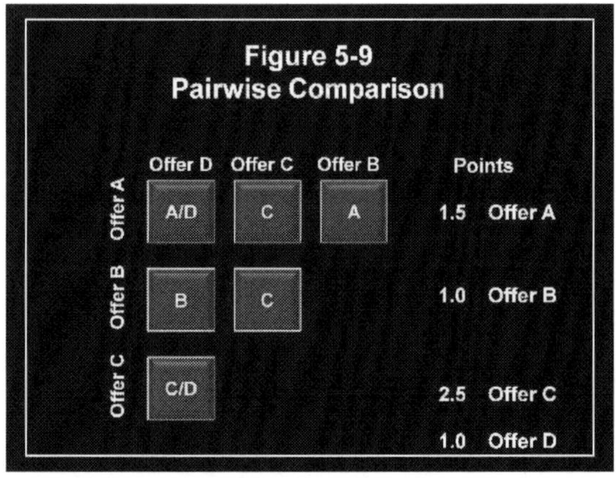

Figure 5-9 is arranged such that all of the pairwise comparisons can be made. First, let's compare Offer A with Offer D. The applicant decides it is a tie (expressed as A/D), so give ½ point to Offer A and one-half point to Offer D. Next, look at Offer A versus Offer C. Offer C wins and gets one point. When Offer A is compared to Offer B, Offer A wins and gets a point. Now go to Offer B on the left. Offer B versus Offer D is won by Offer B. However, Offer C wins over Offer B. Finally, Offer C versus Offer D is a tie. Now we have compared every possible combination in Figure 5-9.

The awarded points are listed on the right.

Offer A gets 1.5 points (Offer A versus Offer D is a tie, for ½ point, plus a win against Offer B for one point. Total = 1.5 points.

Offer B's only win was against Offer D, for one point.

Offer C gets 2.5 points (a tie with Offer D, and wins over Offers A and B).

Offer D only gained two ties for one point.

As a result, Offer C is the clear winner based upon pairwise comparison.

You can see that pairwise comparison is a quick, not very detailed, look at a decision amongst alternatives.

You cannot take into consideration multiple criteria. This process requires a quick, informed judgment call. Once again, this is fine when reducing twenty alternatives to 5, not good if you want to go from 5 to one.

The Simplified Additive Model

When we make decisions, we decide based upon what we care about; in other words, we decide consistent with our beliefs or values. Then we use facts or data to justify our decision.

If you don't believe in the data, you may reshape or reinterpret the data to fit your values. Politicians are masters of this, and do this to get elected.

Why choose between alternatives or available options rather than let whatever happen? The answer is, we are concerned about the consequences of the choices: either avoid undesirable consequences or achieve desirable ones. The relative desirability of these consequences is a concept based upon values.

Decision analysis is an excellent way to make a decision when value judgments must be made. For example, if you are making a decision solely based upon cost, then decision analysis is not appropriate because a value judgment is not necessary. However, if you must make a selection that involves various policy options (reduce air pollution versus imposing a tax on emitters) or when there are multiple conflicting objectives (cost versus environmental impacts), there is no better way than decision analysis to solve a problem.

Stakeholders will have an important view about values. As decision analysts, we need to allow stakeholders to express their value judgments and explicitly include them in the decision process. The technical team (consultant) becomes responsible for acquiring the data and interpreting the data because they are the experts regarding the data.

A client's decision should be based on two types of information: a) the possible impacts of selecting each project, and b) the values the client used in evaluating these impacts.

The first category includes information about the environmental, economic, and constructability impacts of each project. This information is gathered and assessed by disciplinary experts (e.g. environmental scientists, engineers, cost estimators, etc.) and professional judgments made where needed.

The second category includes value tradeoffs among impacts (i.e., assigning relative importance weights to the criteria), which must be elicited from the stakeholders (i.e. the client) because it is their decision to make. We use their value judgments. *If the consultant assigns these weights, the decision is made based upon the consultant's values, which will be different (and often incorrect).*

Yet, most consultants will propose to implement a selection or decision process and return with an answer to the question "which option or project should be selected"? Their assumptions and value judgments likely will be implicit and unstated (that is, hidden from view). So, the client can usually *see* what was done, but not *why* it was done.

Decision analysis improves the quality of the decisions in many ways, but most importantly, by involving the client directly in the decision process. In addition, the public can be included directly in the decision process, which encourages participation because the public actually sees that they have some influence in the process. This is particularly useful when there is a Citizen's Advisory Group or similar public group that provides input to a project. Finally, a decision analysis approach will provide the best alternative and this selection will stand up to scrutiny.

A decision model we call the Simplified Additive Model (SAM) will be described. I have used this model hundreds of times to perform energy site selection, technology choice (e.g. carbon sequestration technologies, solar energy technologies), public policy alternatives, siting of industrial facilities, siting of landfills and waste incineration systems, and strategic plan analysis.

SAM and its decision rules were derived from the Additive Utility Function or the Weighted Sum Model see, *Siting Energy Facilities*, (Keeney, 1982) or *Strategic Decision Making* (Kirkwood, 1997). A number of decision rules are used in this model to provide a sound technical foundation without introducing the mathematical complexity of multi-attribute utility functions.

It is ideal for solving decision problems with multiple objectives or goals that can conflict with one another. For example, how do you balance cost, environmental, and safety issues?

The use of SAM requires that a finite set of alternatives has been identified, and each alternative is represented by its performance against multiple evaluation or decision criteria. Each of the criteria can be weighted differently, and the criteria can use different units of measure. Also, probabilities can be used to address uncertain data.

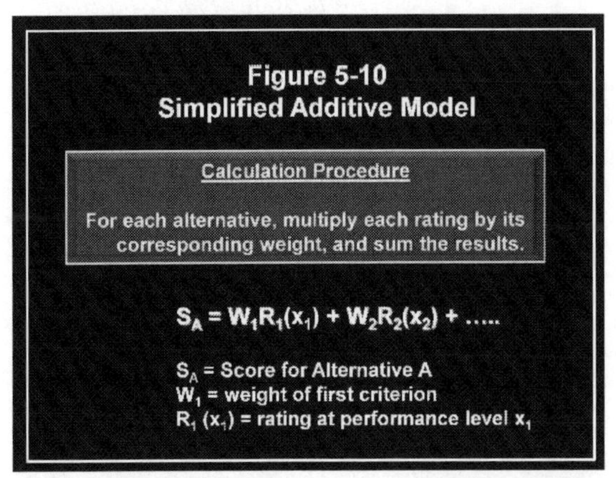
Figure 5-10
Simplified Additive Model

The Simplified Additive Model is shown in Figure 5-10.

For each alternative, multiply each rating by its corresponding weight, and sum the results, where W is a criterion weight, and R(x) is a rating at performance level x. SAM can be applied by following the steps shown in Figure 5-11.

Each of the five steps shown in Figure 5-11 are described below:

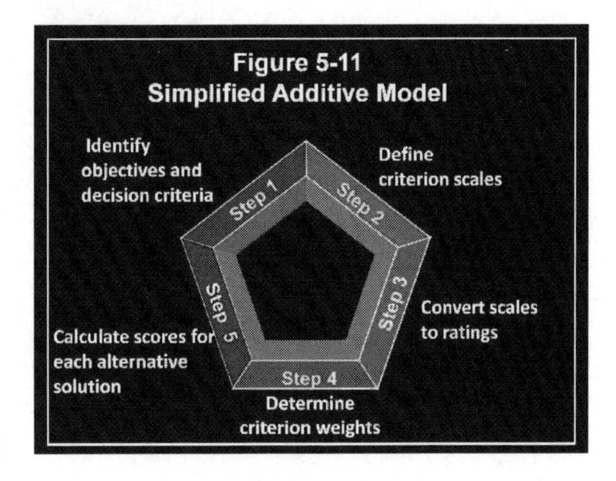
Figure 5-11
Simplified Additive Model

1. Identify the project objectives and the appropriate decision criteria. This step should be conducted with input from the stakeholders because we are dealing with the value structure. They are the decision-makers, and

their values are more relevant than the consultant's values regarding the decision problem.

2. Create measurement scales, or performance levels, for the criteria, and locate the alternative solutions on the scales. This step is lead by the consultant because a variety of discipline experts are required to address the various evaluation criteria.
3. Convert the scales to "ratings". We'll see how this is done in a moment. Again, this is a task for the consultant.
4. Assign importance weights to the criteria. This step should be performed by the stakeholders, with the consultant acting as a facilitator. Setting weights is a value decision.
5. Calculate the scores for each of the alternatives using the equation shown on Figure 5-10. The final step is calculational, and is conducted by the consultant.

We will use examples to further elucidate how SAM is implemented. Each of the five aforementioned steps is further discussed below.

Step 1: ID Objectives and Criteria. In the first step, we identify the project objectives, create an Objectives Hierarchy, and define decision criteria, also called performance criteria. We discussed how to develop objectives in Section vii of this chapter. We'll begin here with an Objectives Hierarchy, which will be used to identify criteria. Our example hierarchy is shown in Figure 5-12.

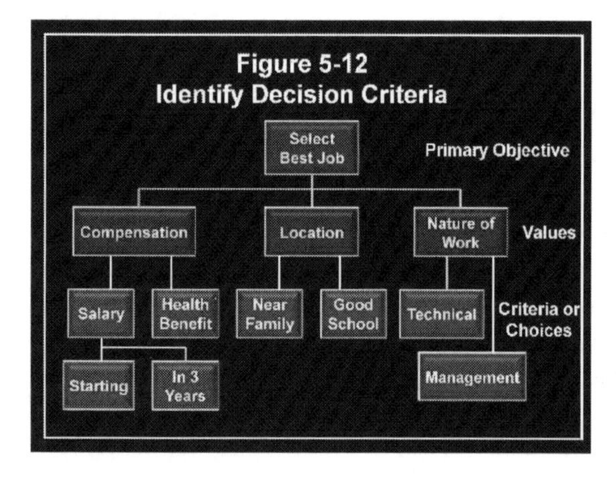

We use objective hierarchies to guide our thinking about a problem. Let's say you are looking for a new job. Perhaps you just moved into a new area. You are married with two school-age children. Finding the best job is your primary (top level or ends) objective. How do you find the best job for you?

The first step is to list your overall objectives. Based upon your values, what is most important to you? After thinking about this, let's say you decide that compensation (salary plus benefits), location, and the nature of the work are most important to you.

The next step is to think about what is important about each objective. For example, when you think about compensation, you specifically mean salary and health benefits. And, with regard to salary, you are interested in not only the starting salary, but your ability to receive increases in the next couple years. With regard to location, you want to be near family and school.

Notice that the items at the bottom of this hierarchy are easier to define and measure. We can measure the distance between the home and the school, for example. We can have a specific goal for salary. On the other hand, a higher-level objective, such as "nature of work" is difficult to define. See what we have done here. We've broken down the top-level objectives into sub-objectives, and then to measurable criteria. Note that these criteria must conform to the following decision rules:

1. The criterion must be differentiable. By that we mean the criteria must vary across alternative solutions. If a criterion is barely different from alternative to alternative, it should not be included because it will not help differentiate among alternatives, and it will bias the result.
2. The criterion must be operational. In other words, you must have the resources available to collect the required data for each criterion and perform the evaluation. In addition, the criterion must describe the possible consequences or impacts with respect to the associated objective. The consequences must be clear.
3. The criterion must be measurable. That is, a criterion that is measurable defines the associated objective in more detail than that provided by the objective alone.
4. The criterion must be non-redundant. That is, each criterion must measure a different issue or concern that is unrelated to the other criteria.
5. The criterion must be independent. That is, if one of the criteria is varied on its scale, it should not change the value of any other criterion. The criteria should be independent of one another.

When you complete the objectives hierarchy, check to see if the criteria satisfy these important rules.

In summary, the objectives hierarchy is a great tool for structuring our thinking, keeping clear what is really important, and defining our decision criteria.

Step 2: Create Performance Levels for Criteria. In Step 2, we develop performance levels, or impact scales, for the criteria. Performance levels (or scales in a continuum), measure the degree to which an objective is achieved. For example, if we want to purchase a new car, and acceleration is a criterion, we might create a scale, or levels, indicating various rates of acceleration from lower to higher. To do this, we will need data on each criterion. If the data are known, creating individual levels, or creating a continuous scale is straightforward. However, if the data is uncertain, simple probabilistic assessment can be made to create impact levels.

In some cases, there may not be any real data. For example, one criterion might be the level of public opposition to a project. Performance levels can be created using professional judgment from experts in public relations, or the results of a public opinion survey can be used. If professional judgment is used to create performance levels, be careful of biases that can affect the scale. It is best to use multiple sources.

When we create performance levels, the following rules must be followed:

1. Performance levels, or impact scales, must be based upon actual data, not scales like "low-medium-high" or "1-2-3". If there is a need for a subjective scale, each level must be clearly defined (we provide an example below).
2. There must be alternatives in the best and worst levels. In other words, the best and worst levels cannot be "empty". They must be populated with data from alternatives. Leaving empty levels at the top or bottom of the scale can introduce a bias in the result.
3. The intervals between each level on a scale should be relatively equal. If a scale is strongly skewed in one direction, weighting the criteria becomes more complex. Most of the time, creating equal scale divisions will not be a problem.

4. All levels must go in the same direction, either low to high or high to low. Either convention is acceptable, although I prefer to use low to high because it is less confusing.
5. As a guide, most criteria should have 3-10 performance levels, based upon complexity and detail available.

Let's go to an example. Figure 5-13 shows performance levels created for a criterion called "Impact on Wetland Habitat" that we used in a site selection study.

Figure 5-13
Performance Levels

Impact on Wetland Habitat

1. < 1 acre
2. 1-5 acres
3. 5-10 acres
4. 10-15 acres

The first performance level is "less than one acre". The next performance level 1-5 acres, followed by 5-10 acres, and finally 10-15 acres. *As mentioned above, the levels must be consistent with the actual data for wetlands impact.* For example, if one alternative impacted wetlands by 16 acres, we would have to adjust the levels accordingly.

The intervals between each level should be equal or roughly equal, making the scale linear. There must be alternatives in the best and worst levels, and many, but not necessarily all, of the other levels. If the alternatives crowd into one level, redefine the levels.

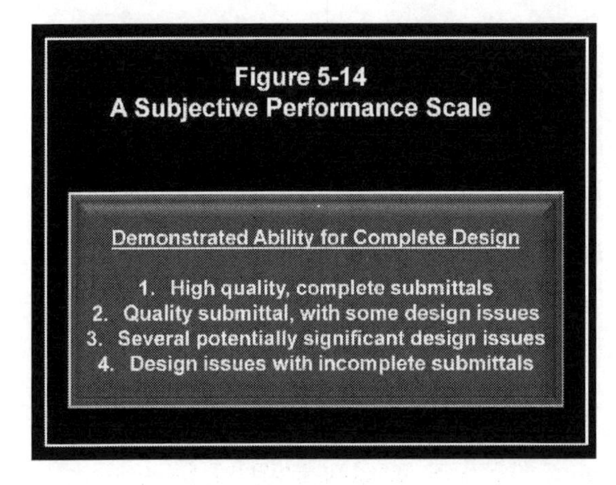

In some cases, you will not have specific numbers for the performance levels, and a subjective scale will be used. Figure 5-14 shows a subjective scale called "Demonstrated Ability for Complete Design Submittal". This criterion was used in a technology choice project where we ranked waste conversion technologies such as gasification and anaerobic digestion

The concern was the ability of the technology vendor to provide a complete design. Conversion technologies include pre-treatment facilities, a processing unit, an energy recovery system, and environmental controls. We were seeking a vendor with capabilities in all of these systems.

We have four performance levels, shown in the slide. These are shortened descriptions to fit on a slide. The key point is that you can create subjective scales like this, so long as each level is specifically defined, understandable, and based upon the available data from the alternatives.

Step 3: Convert Performance Levels to Ratings. Next, we must express the measurement scales in terms of a measure that puts all of the criteria on a consistent basis. We assign what we call "ratings" to the performance levels because criteria are measured differently: cost, miles per gallon, time, losses of wetlands, etc.

We have two additional rules here:

1. We assign 100 points to the best performance level for a criterion, and 0 points to the worst level for that criterion.
2. We assign proportionate points to the intermediate levels.

Figure 5-15 shows how we added ratings to the Impact on Wetland Habitat criterion.

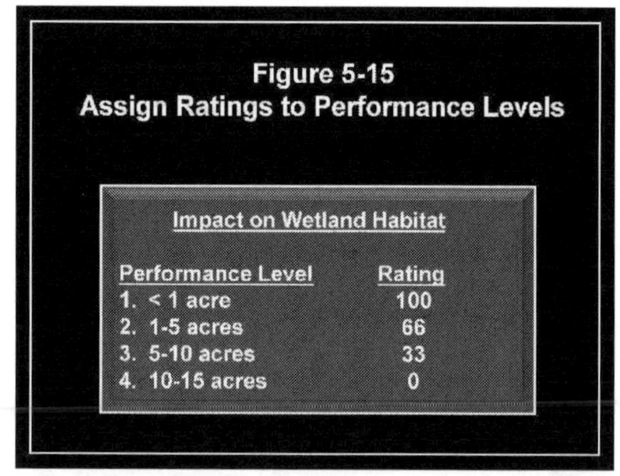

From Figure 5-15, the best performance level, <1 acre, was assigned a rating of 100, per the rule. The worst performance level, 10-15 acres, was assigned a rating of 0, per the rule. The intermediate levels were assigned linearly proportionate ratings of 66 and 33.

Step 4: Assign Importance Weights. It is unlikely that all the decision criteria will be of equal importance. Therefore, we'll need to add weights to our criteria to reflect the differences in importance among criteria, or in other words, to recognize the preferences for each consequence or impact.

We have one additional rule. Weights are a function of both *the intrinsic value and the range of the criterion scale.* Intrinsic value is the importance you would typically assign to the criterion. The range is the interval between the highest and lowest values on the performance scale for the criterion.

Let's revisit the wetlands habitat impact criterion. Suppose the range of impacts was 0.5 to 1 acre. In this case, even though impacting wetlands is important because there are protections against removing wetlands, the amount appears to be relatively small. Mitigating less than one acre of wetlands is likely to be affordable via a wetlands bank. But the more important point here is that the differential impact across all sites is very narrow. Therefore, the weight assigned to this criterion should take this into account. No matter what site or project is selected, the wetlands impact will be within ½ an acre. The weight assigned should be correspondingly low.

Now suppose the range of impacts on wetlands is 5-25 acres. There is a more significant difference in wetlands impact across the sites. In other words, the differential impact is more significant. In this case, a higher weight on the wetlands criterion may be indicated than in the first case.

Finally, suppose the wetlands impact was in a range of 20 and 22 acres. That appears to be a significant acreage. However, look at the range, which is only two acres. No matter which project is selected, the differential impact will be less than two acres. Therefore, the weight for this criterion should be relatively low.

Remember, we are searching for the best alternative. Criteria that vary more over the alternatives likely will create more *relative or differential impacts.*

With that introduction, we will discuss how to assign weights to criteria. There are many different techniques in the literature. However, the "swing weight" method is one I've used successfully for many years, and it is relatively easy to perform with client groups. Recall that setting weights is a value decision, so they should be assigned by the stakeholders. We use the siting of an ammonia manufacturing project as an example.

Weights are assigned by following three steps:

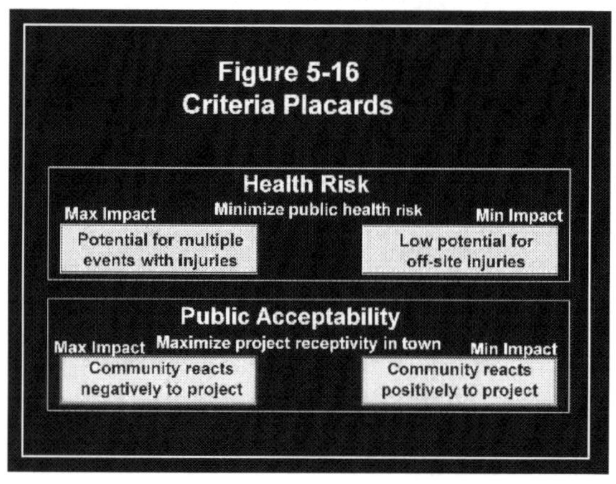

- **Step 1: Rank the criteria by importance.** One ranking method that works quite well is to create placards, either using white display boards cut into 3-inch by six inch placards and make them similar to the ones shown in Figure 5-16. Put some Velcro on the back. Each placard contains the highest or maximum impact performance level and the lowest or minimum performance level. The stakeholders see both the criterion as well as the range. Show all the placards to the group and ask them to indicate the most important criterion. Put that placard at the top on a wall. Then ask which criterion is the least important. Put that placard at the bottom. Next, ask which criterion is second most important, then the penultimate in importance. Use this process to rank all the criteria. The ranking is shown in Figure 5-17.

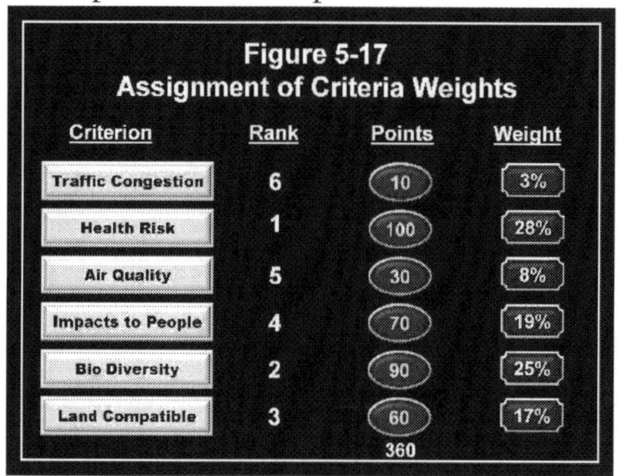

Figure 5-17
Assignment of Criteria Weights

Criterion	Rank	Points	Weight
Traffic Congestion	6	10	3%
Health Risk	1	100	28%
Air Quality	5	30	8%
Impacts to People	4	70	19%
Bio Diversity	2	90	25%
Land Compatible	3	60	17%
		360	

- **Step 2: Ask the stakeholders to assign points to each criterion.** The purpose of this exercise was to establish the "spread" between each criterion. In other words, how close in importance was Bio Diversity to Health Risk? This was done by assigning zero points to a baseline condition, in which all of the criteria are set at their minimum levels. Then points were assigned to each of the criteria, with 100 points assigned to the most important criterion (Health

Risk), and lower points assigned to the other criteria proportionate to their relative weights. You can see that the number of points was relatively consistent with the rank order. However, Impacts to People was judged to be slightly more important than originally thought, as shown by its higher point score than third-ranked Land Compatibility. The total of the ratings assigned was 360, as shown at the bottom of the Points column.

- **Step 3: Calculate weights by dividing each point score by 360.** For example, the Traffic Congestion criterion point score is 10, and 10/360 or .03 or 3%. As you can see, Health Risk received the highest weight at 28% out of 100%.

- **Step 4: Calculate Scores for Alternatives.** In the final step, the scores for the alternatives are calculated. The best Job Offer example is resurrected for this illustration. The data are shown in Figure 5-18. There are three job offers, A, B and C. The criteria are Salary, Vacation Time, Commute Time, and School Quality. The salary offers ranged from $75,000 annually to $95,000. Vacation time ranged from none to four weeks. Commute time ranged from five minutes to fifty minutes, and school quality was best, good, and worst (We do not recommend using these general levels. They are used here for simplicity).

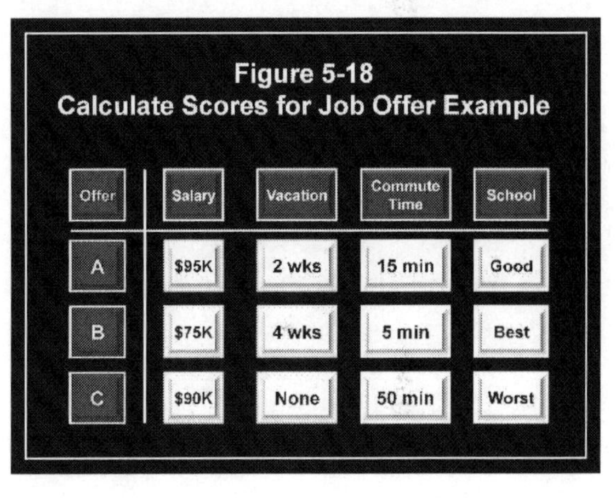
Figure 5-18
Calculate Scores for Job Offer Example

Offer	Salary	Vacation	Commute Time	School
A	$95K	2 wks	15 min	Good
B	$75K	4 wks	5 min	Best
C	$90K	None	50 min	Worst

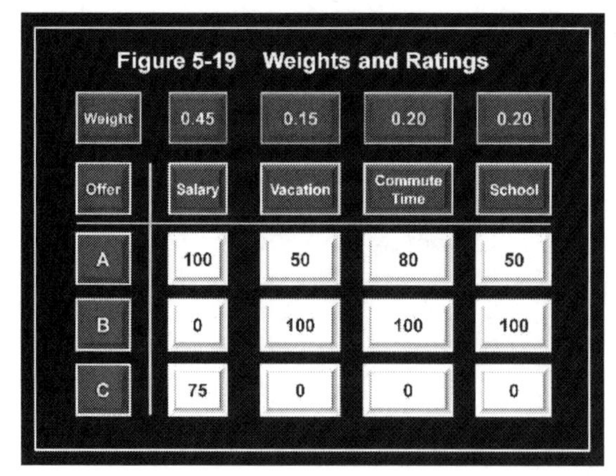

Figure 5-19 Weights and Ratings

Weight	0.45	0.15	0.20	0.20
Offer	Salary	Vacation	Commute Time	School
A	100	50	80	50
B	0	100	100	100
C	75	0	0	0

Figure 5-19 shows the weights and the ratings. The ratings are based upon the data in Figure 5-18.

Let's start with the Salary criterion. The weight of this criterion is 0.45. Then, 100 points was assigned to the highest salary, $95,000, and 0 points to the lowest salary, $75,000. Seventy-five points was assigned to the $90,000 salary because it is 25% or ¼ of the way from $95,000 to $75,000, assuming a linear relationship with salary.

With regard to vacation time, we assume a linear relationship, i.e. 50 points is awarded to 2 weeks' vacation, with 100 points for four weeks, and zero points for no vacation. The weight of the vacation criterion is 0.15.

The Commute Time criterion is linear with time, so Offer B got 100 points with a 5-minute commute, Offer C got 0 points for the 50-minute commute, and Offer A received 80 points with a commute of 15 minutes.

The school criterion received 100 points for the best school, and 0 points for the worst school, and 50 points for the good school.

Finally, there is an intermediate calculation, where the weight was multiplied with each rating for each criterion. For example, for Offer A and Salary, we multiply a rating of 100 points by the .45 weight to get 45 points. The 45 points is inserted into the box that had 100 in Figure 5-19. Do the same calculation for all twelve boxes in the figure. Then simply add the scores for each Offer. The results are:

Job Offer A = 45+7.5+16+10 = 78.5
Job Offer B = 0+15+20+20 = 55
Job Offer C = 34+0+0 = 34

The job seeker should accept Offer A. You will find the Simplified Additive Model to be a very useful technique for selecting from amongst several alternative solutions.

Comparing Subjective Criteria Versus Cost

Up to now, we've considered only situations where you need to select the best option, alternative or project, based upon a set of criteria that typically does not involve cost (earlier, we discussed decision tools that only consider cost). However, when cost criteria must be compared with non-cost, or subjective criteria (for example, environmental criteria), this trade-off becomes more difficult to consider. One way to handle this situation is to convert the subjective criteria into "dollar equivalents". This is complicated. There is another way!

In some decision situations, such as deciding which site is best for a new manufacturing facility, or selecting the best project to advance, we are faced with a difficult problem: we have several non-cost decision criteria, and cost is a significant issue. In addition, the client has engineers who worry more about cost, and an environmental department that is concerned about other issues. How do you make a decision in this situation? If you apply the Simplified Additive Model, you will find that cost issues likely will dominate because of the importance of cost, so you've built in a large bias into the analysis.

One solution is to separate cost from all of the non-cost issues. The non-cost issues are rolled up into one measure, for example, "environmental-infrastructure". If we are looking at the cost of several alternative projects, the cost is presented as a differential cost, using the lowest cost project as a baseline.

A tool we use to separate non-cost and cost issues is an "opportunity frontier scatterplot". An example of an opportunity frontier scatterplot is shown in

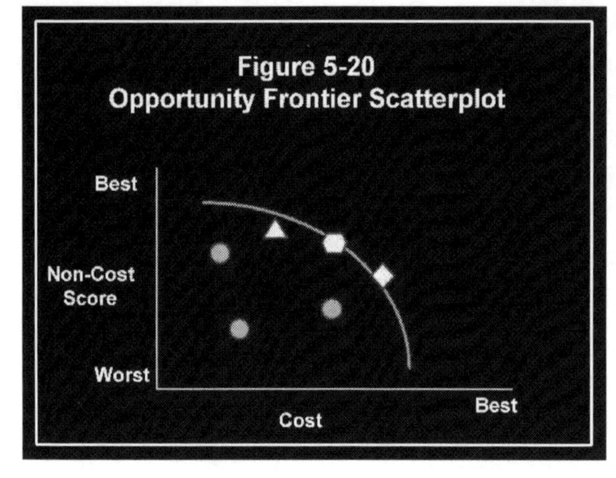

Figure 5-20. Non-cost issues, rolled up into one measure, are plotted against cost. The alternatives are shown as circles, a hexagon, a diamond, and a triangle.

The highest non-cost score will be in the northern portion of the plot. The best cost will be in the southern portion of the plot. However, the alternatives with the best score on non-cost issues and the best cost will be in the northeast part of the plot. This is the "frontier". The alternative using the hexagon shape would be the best alternative if the stakeholders equally weighted non-cost and cost issues in their deliberations of which alternative is best. On the other hand, if the stakeholders thought that cost should be weighted somewhat higher, they might choose the triangle alternative. Or, if the stakeholders thought that non-cost issues were more important than cost, they might select the diamond alternative. So, you can see that this scatterplot makes a trade-off of non-cost versus cost issues quite straightforward, while keeping the final selection close to the frontier, which is where the best alternatives are located. In my experience, this approach works well before a client team because they can decide what is more important: cost or non-cost issues.

Let's look at an example. Figure 5-21 shows an opportunity frontier scatterplot for a power generation project site selection study. Site differential cost is plotted against a rolled up environmental-infrastructure score that includes various

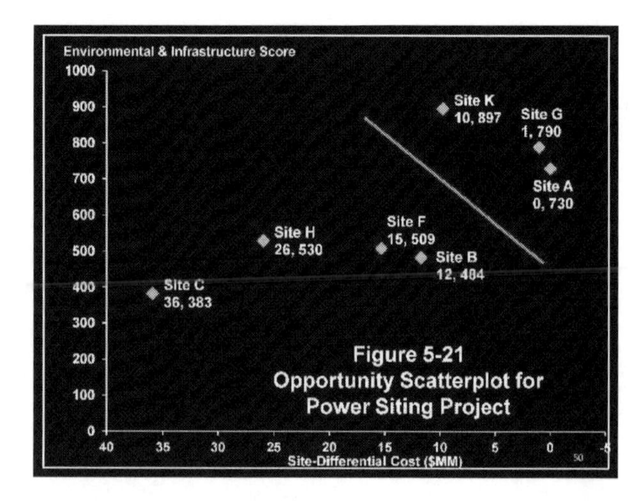

environmental and infrastructure issues that could not be easily converted to a cost.

The first step, after plotting the sites, is to define the "frontier", or the sites that have the best combined cost and non-cost score. In this plot, we drew the straight line, showing that three sites were "on the frontier", Site K, Site G, and Site A. These sites are clearly better than the others.

Next, your client gets to decide which plant site is best for them. You facilitate the decision, but the decision is theirs. In this case, they have three alternatives that are close. They select Site G to maximize on non-cost and cost criteria. They will Select Site A if they think cost is a bit more important than non-cost criteria, or select Site K if environmental issues appear to be quite important.

Once again, I've used this tool many times. It is excellent for client presentations. Also, it works well for public presentations, where the public is typically more worried about the non-cost issues, while the client is more concerned about the cost issues.

Afterword

About the Author

Dan Predpall, P.E. is the Founder and President of Successful Consultant Training LLC.

- B.S. Physics and Electrical Engineering, Stevens Institute of Technology
- M.S. Physical Oceanography, Long Island University
- M.B.A. Quantitative Analysis, New York University
- Key Mentor: Dr. W. Edwards Deming
- Former Vice President, URS Corporation (now AECOM)
- Former Sr. Vice President, Pacific Valley LLC (solar development)
- President, Opti-Site International
- President, Southeast Solar & Power LLC

One of the most important things that ever happened to me began, improbably, in a statistics class. The course was taught by Dr. W. Edwards Deming, one of the most remarkable men of the 20th century. When he became my mentor, he was one of the most influential and successful consultants in the world, having, among many triumphs, dramatically revolutionized the manufacturing cultures in both Japan and America. What I learned from him, gradually over a number of years, changed the direction of my life. His kindness, generosity, and the thousands of lives he helped transform for the better continue to inspire me. After 40 years of consulting, I want share what I've learned, to give back by making a positive difference in the lives and careers of other individuals and consultants. In addition, this training can make a positive impact on consulting firms as well.

One of the principles Dr. Deming emphasized over and over is the importance of a commitment to "continuous learning." I like to say it like this: "continuous learning is the best path to prosperity." One thing I've loved about consulting is that the learning never stops. Consulting has provided me with success, flexibility, freedom, and self-respect.

A fine compilation of Deming's management philosophy can be found in the book *The Symphony of Profound Knowledge* by Edward Martin Baker (Baker, 2016).

I couldn't write a book like this without the help of many people throughout my career who believed in me and took the time to teach me new concepts and techniques. To name just a few, Dr. W. Edwards Deming, Dr. Ralph Keeney, Dr. John Lathrop, Adrian Bowden, Dr. Jerry Hollinden, Dr. Melvin Esrig, Dr. Harry Horn, Chuck Wahtola, David Saul, Timothy Cohen, Sally Hogshead, and Brendon Burchard.

About Successful Consultant Training LLC

Successful Consultant Training (SCT) offers training to consulting firms and consultants that covers the four key skills and knowledge every consultant should have: business development, client service strategies, problem solving and decision-making, and a consultant's mindset.

The Desk Reference for Training

In the Foreword, we suggested the use of this book as a training vehicle. This can be accomplished by simply using the book as a reference. Or, management can utilize the book for conducting in-house training, as described in the Foreword.

Live Online Training Webinars

Successful Consultant Training LLC also offers live webinars that address the four key skills. There are twelve two-hour webinars given live over the Go To Meeting platform. Each webinar comes with a Guidebook filled with exercises, templates and checklists that augment the course material.

The webinars are:

- Business Development and the SCT Compleat Marketing Model
- A Systems Approach to Business Development
- The Psychology of Persuasion
- Preparing Winning Proposals
- Preparing Winning Presentations

- Key Client Strategies
- Client Service and Delivery
- Innovation and Project Management
- Problem Solving
- Decision-making
- The Consultant's Mindset
- The Fascination Difference Advantage System

About the Fascination Difference Advantage System

Chapter 3 includes a discussion of Fascination, a unique approach to communication and marketing. For all of us, it is vital that we learn how to communicate effectively with prospects and clients in a way that is persuasive and influential. Fortunately, we have a built-in authentic communication style that can accomplish this goal. However, that style may be hidden or partially hidden due to previous training.

The Fascination Difference Advantage System is an adaptation to consulting of the Fascination Advantage™ program developed by Sally Hogshead of How to Fascinate. I am a Certified Advisor of Fascination, and I was trained by Sally.

You can learn to communicate effectively by getting a Fascinate Communication Profile. To get a Profile, you complete a short questionnaire online and immediately receive your Profile. The Profile provides an in-depth analysis of how your communication is received by others. The Profile shows you how to unlock that built-in persuasive communication style and use it in both oral and written form to influence and persuade prospects and clients.

As described in Chapter 3, there are a number of other uses of the Profile, such as making better hiring decisions and conducting more effective performance appraisals that may be of interest to managers.

If you are interested in getting the Profile, go to the following website:

https://ea106.isrefer.com/go/FAA/ConsultingFirms/

There will be a fee for getting the Profile.

We offer a 90-minute webinar that shows staff and managers how to implement the Fascination Difference Advantage System in your office.

For more information about Successful Consultant Training LLC's professional development programs, call us at 805-451-7658, email us at dan@successfulconsultanttraining.com, or visit www.sctprofessionaldevelopment.com

Appendix A: Preparing Winning Proposals and Presentations

So, what makes a winning proposal or presentation? A winning proposal or presentation is just a well-written sales pitch! Winning proposals or presentations is *all about selling*, not describing your technical prowess, or providing a price quote, or writing a nice scope of work, or presenting a complete company history.

Preparing Winning Proposals

What are the goals of a proposal?

- Win on the basis of value rather than price. Where is that value proposition we talked about in Chapter 6?
- Offer a customized solution tailored to the needs of the prospect. Proposals and solutions that look like boilerplate never win. Make the proposal look like it was completely customized for the prospect.
- Make the proposal *persuasive*. We provided many ways to persuade and influence in Chapter 7. Use them!
- Demonstrate that you can write a professional document that is well written and wants to be read. Your proposal must communicate clearly, and be structured in the proper order.

Proposal Success Equation

After writing proposals for decades, it became clear that a few issues were closely linked to the outcome (win or lose). I expressed these issues in the form of an equation: The Proposal Success equation.

Proposal Success = Attention + Value + Educate + Persuade – Anxiety

Attention means get the attention of the reader (prospect). Value means demonstrate value through a value proposition, educate means educate the prospect about your solution and its benefits, persuade means use persuasion techniques to influence the prospect to buy from you, and anxiety means reduce purchase anxiety

(risk).

Now, all you need to do is address these five terms and you will win! Well, unfortunately, it isn't quite that simple. However, based on my experience, attending to these issues will significantly improve the probability of success. Each of the terms of the Proposal Success Equation is discussed below.

Attention

You must find a way to get the attention of the prospect (reader) and then get them to read your proposal! *Here is a startling fact*: when the prospect grabs your proposal to read, it is likely the prospect is not fully focused, or energized, on your proposal at the start. So, the challenge is, *how do you engage the reader?*

Here are several ways to gain the attention of the reader:

- **Tell a compelling story.** This is a great way to start a letter or a section in a proposal. Here is an example story: When I am proposing a large site selection study, I begin the proposal with a story of how I conducted the first phase of a study to locate a gas turbine plant site in Arizona. The client wanted to see the site. We made a visit. The client asked a simple question when we got to the site. Where is the transmission line? No line was in sight. I did not realize that there was a bust in the GIS database for transmission. Then I went on to say how we eliminate problems like this. A story like this creates images in the prospects mind of embarrassment and concern. We alleviate this concern with our solution. The emotion is fear of failure, loss of valuable time, fear to go to his boss with this result.
- **Ask a situational question.** Start questions like the following: "If this happened, what would you do?"; "Have you ever…."; "How many times have you…."; Has this ever happened to you...." "Do you realize that…". For example, one service I have provided is developing a risk profile for operating facilities that may have hazardous waste issues. In my proposal, or meeting

with the client, I always ask if they are prepared to address these issues. Do they have sufficient insurance to cover the liabilities? These questions expose the fact that the firm does not know if their insurance would cover any liabilities, which gets them interested in my service.

- **Use a catchy photo or graphic.** I recently performed a renewable energy study for a small city. I began the proposal with a high quality aerial photo with a wind turbine and biomass processing unit simulated into the photo. This simulation got the attention of the client, both because the quality of the photo was very good, they got an opportunity to see how their project would look on their landfill site, and they saw that a solution of this kind was feasible.
- **Use a surprising fact that the reader didn't expect to see.** An unexpected fact raises curiosity. When we wrote proposals to determine the liabilities at contaminated properties, we would point out that liabilities with the lowest probabilities may be the biggest problem, which can be contrary to intuition. This statement usually caught the prospect's attention, prompting more questions (which are good).
- **Use headlines in your** text (such as paragraph headlines). Use attention-getting words like "discover", "proven", "quick", "results", "least cost", or "easy".
- **Use attention-getting ideas in a cover letter to the proposal.** Start the letter with a thought-provoking question or the cost implications of your offering that will evoke an emotional response. Significant benefits of your offering also can surprise, and should be front and center in a proposal cover letter.

There must be a strong opening in a proposal that makes the reader want to continue reading.

Value

As we discussed earlier, value must be related to a business driver or priority. What is the most critical outcome the prospect is looking for? Value is a multidimensional concept, as depicted in Figure A-1.

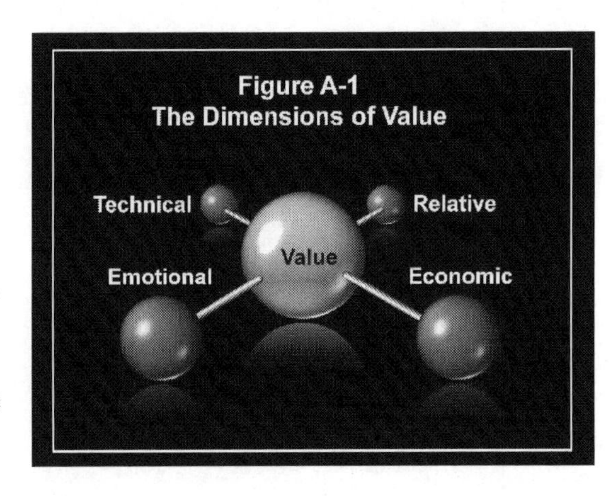

Figure A-1
The Dimensions of Value

These dimensions of value should be considered to see if they are relevant for your proposal. Each dimension is discussed below.

- **Technical.** How does your offering or solution help the prospect? In the technical area, this could be higher quality (increasing reliability, reducing failure rates, reducing customer complaints, using quality in marketing), an innovative solution (a new approach to a problem), a new technology that improves a system, or a solution that enhances a prospect's competitive position. When we give a presentation on site selection to a client, we tell them that our site selection process optimizes the project, which results in the best site available, not just an acceptable site. First, saying this always brings up the question "how do we know it is the best site?" which opens an opportunity to talk more about our innovative solution. Second, we describe how our siting process literally evaluates every land parcel in the study area, and, therefore, we arrive at the best site.
- **Economic.** Relate the value of your offering to an impact on the client firm. For example, you could say "Allow us to get you into compliance with these regulations now, at our low fee of $15,000 and avoid up to $100,000 of assessed penalties." Here you have associated your offer with a key business priority…avoid penalties….and you have compared your fee with a much higher value of your offering. That is a powerful statement. Other economic outcomes include lower costs, reduced expenses,

improved cash flow, increased revenues, and reduced risk. Risk can be a very important issue. Managing liabilities properly almost always costs less than the alternative.

- **Emotional.** What is the prospect worried about, or frustrated with, or angry about? Examples could be increase morale, provide additional staff training, create a better public image, or improve safety procedures. Outcomes in this category could relate to unstated or overt issues. Many of my clients in the site selection business need someone to find a suitable site for a new facility; however, they are very concerned about the public's reaction to the project. Will they accept the results? Will they oppose the project at the permitting stage? Will they get harmful publicity? I tell them about our site selection process, which actually includes the public in the siting process. This almost guarantees acceptance of the site because the public had a role in the selection.
- **Relative.** Value is always relative to something else. Prospects don't assess the value of an offering in isolation. They always consider it relative to other actions, including doing nothing. These actions may include alternative solutions. If you can determine these alternatives, you will be able to focus on differentiation relative to these options, rather than talking about points of parity, which will cloud your value proposition.

Consider each of these dimensions of value when writing your proposal.

Educate

Use education to prove the value you're providing to the prospect. Explain what you do, why it works, when it will work, and how it will benefit your prospect. Present facts and data to support your solution. Facts increase confidence and reduce anxiety. Include endorsements or testimonials that support your solution. Third party support or third party data is excellent. Or, show a case history where your solution solved the same or similar problem your prospect has encountered. Show that you are the expert in this area. This will indicate that you are an authority. People don't question authorities. Remember, when you describe your solution in detail, you differentiate yourself from your competition. In addition, when you educate your prospect about various solution alternatives, and why you chose the one in the proposal, you eliminate much of your competition!

Persuade

In Chapter 3 we presented a number of ways to persuade and influence. With regard to proposals, the most important considerations that drive persuasion are:

1. Increase Desire
2. Increase Aspiration
3. Use Psychological Triggers
4. Pay Attention to Message Sequencing

Each of these is discussed below.

First, we want to increase desire for our offering. We discussed how to do this in Chapter 3. For a proposal, use a "results in advance" approach, where you show the prospect what the future will look like with your solution implemented. Then show proof that this solution works. This creates positive expectancy. If your cost will be viewed as high, provide justification for the cost. If the client specifies proposal evaluation criteria, exceed them.

Second, we want to increase aspiration, also discussed in Chapter 3. For proposals, reducing anxiety (purchase risk) is vitally important. Demonstrate that you will increase certainty. Show that there will be no failures. You have anticipated every situation or event. How will the prospect gain a benefit? Will he get recognition? Will he achieve a professional or personal goal?

Third, apply several psychological triggers presented in Chapter 7.

Fourth, how you structure the proposal will affect how persuasive it is. The order that you present information, or topics, makes a difference with regard to the persuasive power of the document. The outline that is most often used is:

I. Introduction
II. Background/Problem Understanding
III. Scope of Work
IV. Methodology
V. Project Organization/Project Team
VI. Qualifications
VII. Schedule and Cost

Next, here is a proposal outline that tells a story:

I. An attention-getting statement
II. Present the key problem/frustration/concern
III. Reinforce impact of the problem with 3rd party data
IV. Prove your solution works.
V. Replace pain with gain.
VI. Project Team
VII. Schedule and Cost

Here is an outline typically called Results in Advance. The idea is that the prospect will see himself achieving the result you just presented to him.

I. Describe where the prospect is today
II. Describe where the prospect wants to be
III. Show a 4-8 step process to get there
IV. Demonstrate the challenges along the path
V. Show how you will overcome them
VI. Project Team
VII. Qualifications
VIII. Schedule and Cost

Finally, here is another variation of the previous two outlines:

VI. The Promise

VII. The Challenges

VIII. Case History

IX. Solution Overview

X. The Proposal

XI. The Benefits

XII. The Fee

XIII. Risk Reducers

XIV. Project Team and Qualifications

The three outlines shown above have been used successfully many times. They are shown to provide ideas regarding proposal outline alternatives, including the order of topics in the document. Once again, the sequence does impact the persuasiveness of the document.

Reduce Anxiety

The fifth and last term in the Proposal Success Equation is anxiety, or purchase risk. Every time a prospect or client considers a purchase from a consultant, there is the element of risk, as translated into fear of failure or loss, loss of confidence by his/her boss, making a mistake, or similar concern. It is best to address this risk in your proposal.

There are two types of purchase risk: rational risk, based on logic, and emotional risk. Rational risk is what you would expect. For example, will the promised results be realized? Will the budget be met? Will the project meet the tight schedule?

The emotional risk is more complex. Here, the primary concern on the part of the prospect is the fear of loss. Psychologists have determined that people always fear a loss. It is natural. Emotional risk can be reduced by using a testimonial, for example, where a trusted associate speaks glowingly about your proposed solution.

Here are several more ways to reduce the risk of purchase:

1. A project manager can alleviate some risk by spending more time at the prospect's location during the first days of the project.
2. Delay the purchase. Perhaps having a bit more time would allow for gathering more data to support the decision.
3. Emphasize the success of your offering using case histories. Examples that are relevant to the opportunity are best. Emphasize the benefits gained by the client. Mention any staff who worked on the case history and are assigned to this one.
4. Provide a guarantee or warranty. This can be difficult, but it is worth mentioning. Get your contracts people thinking about what they can do. Offer to absorb time to redo poor work products, or provide "bonus time" that can be used on the next project.
5. Offer a discount to encourage the prospect to award more work to you.
6. Stress your customer service. Is there something special about how you service your clients? If not, create one!
7. Compare your fee with value offered. Comparisons are psychologically powerful ways to convince a prospect to buy from you.
8. Offer a celebrity expert, who has exceptional qualifications and recognition, if possible. Use the celebrity in a senior review role.

In summary, keep this proposal success equation in mind while preparing your proposal. Be sure you address each term of the equation in the proposal. You will see the benefits in your win ratio!

The Essential Agreements

This topic is not directly addressed in the Proposal Success Equation, but it serves as a good summary.

In the proposal, we want to create a stream of agreements or acceptances that eventually will lead to selection of the proposal by the prospect. You want the prospect to agree with everything you say in the proposal. If you accomplish that, you likely will win.

Here are ten essential agreements you must attain in your proposal:

1. Does the prospect know of your firm, and have a positive view of your firm (i.e. have you positioned the firm ahead of the proposal effort?)?
2. Does your proposal demonstrate a solid understanding of the prospect's problem?
3. Will the prospect understand and appreciate your proposed solution?
4. Will the prospect believe that your solution is a proven one?
5. Are the covert benefits aligned with the prospect's business drivers or business priorities?
6. Have you addressed the prospect's covert (emotional) or unstated needs and wants?
7. Does the fee compare favorably with the value (worth) of the proposed work?
8. Will the prospect view the project team as highly qualified?
9. Is the proposal consistent with the prospect's budget and schedule requirements?
10. Have the prospect's objections been addressed?

If you have solid answers to these questions, you should be well on your way to victory.

Proposal Preparation Process

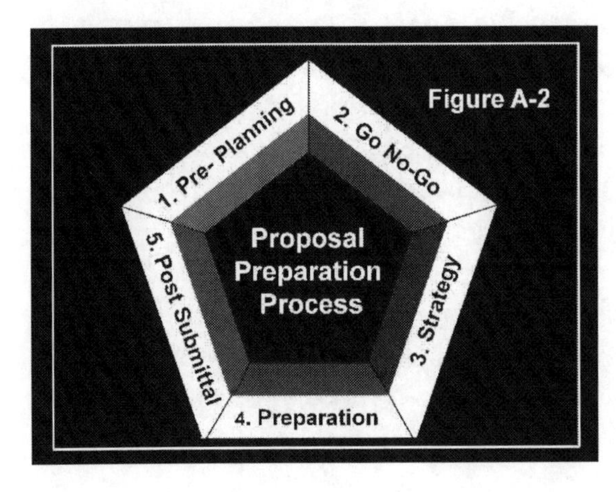

In this section, we will discuss the five-step Proposal Preparation Process shown in Figure A-2. Each of the five steps is discussed below.

<u>*Pre-planning Process*</u>

Prior to Request for Proposal (RFP) release is a key time for intelligence gathering, and this applies to both public and private opportunities. There are a dozen issues that should be researched before replying to the RFP. Identify the decision-makers. Who will make a decision regarding the RFP? Does one person have the authority or a committee?

- Gather intelligence about the problem. Determine what the problem is in detail. Show understanding of the problem. Do they have an ideal solution? Are they open to options? Has a budget been established for the work? Do they have an emphasis on quality or price or time or service?
- Describe your preliminary solution to the prospect, and try to get some feedback. This is critical because you can obtain tacit agreement that your solution was the one they were looking for.
- Who are your competitors? How does your experience and understanding of the problem compare to the competition's strengths and weaknesses? How do you overcome their capabilities? This is a great Go No-Go question as well.
- Does the prospect have a consultant on board now? Who is it? How long has the incumbent been working for this prospect? Is the purpose of this RFP just to get the required three bids before awarding the project to the same consultant (in other words, is the project wired?)? If there is a consultant on board, what can you do to unseat them?
- Gather intelligence about the decision-maker. What are the ambitions, or goals, of the decisionmaker? What is their expectation for results? What type of pain are they frustrated with? What emotional appeal can you make? What are their expectations for consultant performance? What are the key evaluation criteria? Responsiveness? Great looking reports? Cost sensitivity? Schedule? Innovative solution? The right technology? Ability to communicate with the public? Try to find people who are not directly associated with the RFP and ask them for intel. These people could be in other parts of the client firm, vendors who have worked for the client firm, or others who have successfully competed for work with that firm.

- What does the key decision-maker want from you? Security? Quality? Support? Reduced risk? Recognition? His back?
- Help the prospect write the RFP. Suggest that you help the prospect write the RFP.
- Do pre-work in proposal. What task(s) can you conduct prior to submission of the proposal that would show the prospect that you understand the problem and are on the way to solving it. For example, include some modeling, prepare a preliminary CPM chart, or provide photos of the job site with a short discussion of possible issues.
- Visit the job site. It is essential that you visit the job site, if applicable. You will learn a great deal about the problem, and will make your proposal sound much more customized. While there, take photos for inclusion in the proposal. Show a photo on the proposal cover.
- Visit the firm's website. Spend time reading about the firm you are proposing to prior to submittal. View their website. This may provide some insight into their problem or what they have tried in the past.

Attend the bidder's meeting. If the prospect holds a bidder's meeting, be sure to attend. Be careful about asking questions that could divulge your strategy. These meetings also are helpful for finding subcontractors.

Go No-Go Process

The value of the Go No-Go process is to save money, time and effort better spent elsewhere, and as a protection against sending in a poor proposal that ruins your reputation with the client (and, maybe more firms…they talk!). You must determine whether you should hold'em or fold'em!

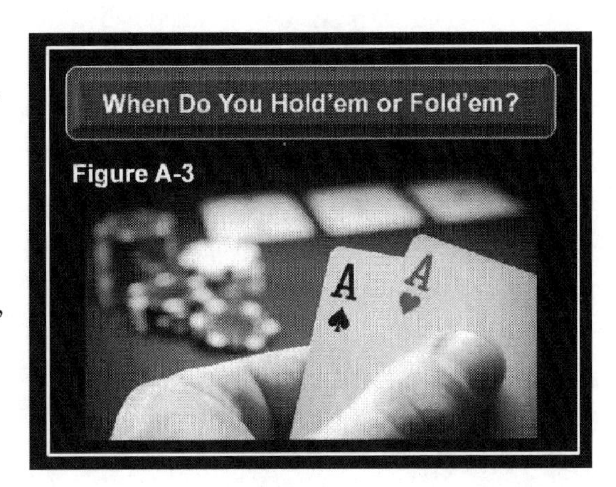

Figure A-3

There are three possible goals of preparing and submitting a proposal. Let's take a look.

- First, you expect to win this important opportunity. The opportunity fits well with your resources and capabilities, and should be profitable.
- Second, you don't think you will win, but you want to educate the client about your firm and its capabilities to position for the future (you only want to do this for a good reason).
- Third, the client expects you to propose, but you don't think you can win. The client needs three proposals. Clearly, you want to avoid these situations as much as possible.

You should fold if one of these concerns applies:

- You have no idea what the budget allocation is for the project.
- You cannot make a profit. This one should be obvious.
- Contract terms are unacceptable. This one also should be obvious!
- You have no real relationship with the firm (of course, you might want to throw in a proposal to start building a relationship, but this is risky because you don't know the prospect's expectations).
- There is an incumbent firmly in place. Is he replaceable?
- There are many competitors, so we have a lottery. Not a good situation.
- You cannot show good compliance with the evaluation criteria. A bad sign!
- You just don't have a solid team, especially a Project Manager who fits well. Be truthful about this one! Don't kid yourself!

  There is no strategic value (it is a pure one-off that offers nothing special). Few companies seem to pay attention to this one.
- You have a conflict of interest with another client.
- Politics will be involved, and you may not be an "insider".

Also, you want to be sure that the opportunity isn't wired to another consultant. Here are five questions you should get answers to before deciding to propose:

- Are the evaluation criteria unusually specific? This might mean they are patterned after the incumbent. Avoid.
- There are no evaluation criteria? If you cannot learn anything about what it will take to win, skip.
- Is the time allowed to respond to the opportunity disturbingly short? A very short time to prepare the proposal is a BIG hint that it's wired.
- Is there an incumbent, and has he been there for a year or more? If so, it may not be worth the trouble and effort unless the prospect strongly encourages you to respond.
- Is the prospect unwilling to give you information about the project? If so, this is an indication that the prospect may have a favorite (and you are not it!).

Above all, the time spent on Go No-Go will be well worth it. You have a limited amount of resources and time. Focus on the best opportunities.

The Proposal Strategy Development

The most effective strategy for winning any proposal will be one where you design the proposal based upon those criteria which your intelligence gathering says the prospect will use to decide which proposal best meets their requirements.

We suggest that your strategy include the following factors:

- Intelligence gathered on the prospect firm and decision-makers
- Internal strategic objectives (what specific results do you want from this opportunity? profit? client relationship?)
- External environment (what is going on in the market, in the economy, and how this is impacting the client)
- Internal resources (what resources can you bring to the opportunity?)

- Strengths and weaknesses (emphasize your strengths and minimize your weaknesses)
- Sub-consultants that will be needed to complete the team.

Here are a few suggested proposal strategies:

- Educate the client about who you are, what you do, what you offer, and how it will help the client;
- Start selling in the cover letter and again throughout the proposal;
- Emphasize your strong client relationships;
- Know how success is measured (i.e. published evaluation criteria? Other criteria from meetings with the client?);
- Include innovative solutions (i.e. what can you do that is different from your competition?);
- Stress client benefits over features;
- Create a readable document, complete with graphics, white space, conversational writing, storytelling;
- Appeal to implicit (overt) as well as explicit needs (the emotional appeal);
- Align your value proposition with the client's buying motives;
- Propose a client-preferred or client-centered solution versus your solution (don't fall in love with your service). Customize your service to fit the prospect's specific needs;
- Clearly understand and restate requirements in RFP. Don't copy them, but add more detail to show deep understanding;
- Consider political considerations;
- Kill your competitors in scope/approach/options that show that your competitor's approach will not work;
- Alternatives are discussed, and eliminated (also can kill off competitors);
- Repeat your compelling theme throughout proposal;
- What are the two or three "hot buttons", the issues or concerns or frustrations that keep your prospect up at night. These hot buttons may or may not be mentioned in the RFP or scope of work. Link the hot buttons to your approach and methodology, so your prospect

knows that you have a deep understanding of their problem or situation;
- Include testimonials.

Prepare the Proposal

In this section we'll cover six important issues regarding proposal preparation: proposal outline, proposal preparation process, work breakdown structure, problem statement, schedule and cost, and red team review.

Proposal Outline. This is probably a good place to start! Recall what we talked about earlier in this section about persuasion in the Proposal Success Equation. The order or sequence of presentation makes a difference. Address the most important issues first and get a series of agreements.

Proposal Preparation Process. This is a process diagram or detailed checklist that shows all of the steps through the proposal preparation process, such as storyboarding, art design, project budget, project schedule, problem statement, client hot buttons, and so on. This process diagram will keep the proposal team on time and will prevent last minute problems. If you don't have a proposal preparation process diagram, sit your proposal team down and ask them to create such a process.

Work Breakdown Structure. Most proposals (except ones with a simple scope) should have a Work Breakdown Structure (WBS). A WBS decomposes the scope of work into smaller tasks. The WBS can be in tabular form or diagrammatic format (see Figure A-5).

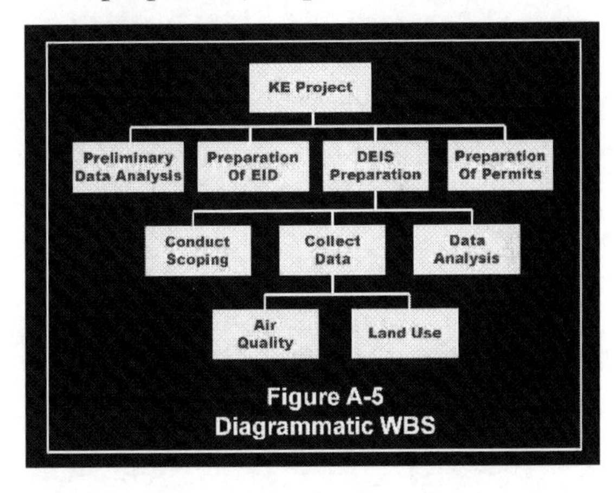

Figure A-5
Diagrammatic WBS

The WBS should be shown in the Scope of Work section, and again in the schedule and budget sections. This will provide a nice logical flow to the proposal. Other advantages of using the WBS include assistance with assigning and tracking responsibilities, and allows for better cost and schedule management.

Problem Statement. The problem statement is a key part of a proposal. Prospects read this section carefully to see if you understand their problem. A well written, clear, and detailed statement will show a prospect that you understand the problem, and that you are an "insider". And, you possibly get a bonus if your competition didn't pay as much attention to this part of the proposal. Describe the impetus for the project (how did it come about?). Has the prospect correctly described the problem? What will happen if the problem isn't solved?

Schedule and Cost. If your billing rate is relatively high, there are ways to justify those rates. Here are a few examples:

- Always try to show a schedule shorter than required by the client.
- A great way is to show the extreme value of your service. Show an important benefit that the client is looking for. Show a benefit that is unique to your team. I was selling a couple of sites suitable for development of 10 MW solar energy projects. I created a 100-page document that looked like an Environmental Impact Report, but in actuality, the data was obtained from brief site visits and online data. However, the perception was that we performed a detailed assessment of the property.
- Uncertain outcomes deserve higher fees. Show the uncertainty. You can show your client how you will reduce that uncertainty. If you do this, you will probably be the only firm who does.
- Try bidding a fixed price or lump-sum contract. Assume the budget risk.
- If your assignment is truly unique, or it requires specialized expertise, you can justify higher rates. One service I offered was a technology selection for sanitary waste combustion or gasification. One of our staff came to us from one of the vendors of this equipment. His knowledge was very desirable, enabling us to get a good fee.
- You may be able to argue that spending more now will save money later. I've used this justification many times when performing early

site assessments, where a small effort can have big returns by identifying risk issues before larger sums are expended at the site.

- Offering more high level time (more experienced people) will more quickly solve the client's problems. Many clients are OK with paying higher rates for truly capable people.
- Compare your price to savings the client will realize, or the increased profit the client will get. Your fee will sound small if you save the client money, or time.

There are ways to reduce your cost as well, such as:

- Can you reduce the amount of field work needed? Can you be more efficient? Can you use more existing data or more analysis?
- Can you use more lower level staff?
- Can you outsource tasks of lower complexity?
- Can you use more technology and less staff time?
- Can you defer some tasks to a later time so that the first phase or task costs less?
- Will economy of scale help reduce costs?
- Can you offer a volume discount?

Red Team Review. The purpose of a Red Team review is to ensure that the proposal is written from the prospect's perspective. Conducting a Red Team Review always improves the clarity and quality of a proposal. It is beneficial to have a different set of eyes read the draft proposal. They will see things that the preparers will not. We strongly suggest that a checklist be circulated to the Red Team to ensure that everyone understands the objectives of the exercise. Ensure that the team reads the RFP and the proposal.

The key tasks for the Red Team are:

- They should ensure compliance with the RFP, and ensure that the proposal indicates compliance clearly.
- They should look for the compelling proposal themes. Red Team members put themselves in the mindset of the client and test the proposal themes.

- Is there sufficient social proof, e.g. case histories, references, testimonials?
- Have you demonstrated clear understanding of the client's objectives?
- How convincing is the value proposition? Does it differentiate the proposal from the competition?
- Is the proposal persuasive? Is it clear and logical? Does it want to be read? Are the first parts geared towards decision-makers so they get the message?
- The drawings and art work clearly support the text, and add information.
- Is the cost justified?

The Red Team review is an important task, must be taken seriously, and deserves the effort required.

Speaking of seriously, try to avoid the following proposal diseases:

- **Motion Sickness**: jumps from topic to topic
- **Senility:** uses to much boilerplate
- **Obesity:** too much information, e.g. fat
- **Sterility:** no original thoughts (no creative solutions)
- **Narcissism:** arrogant, full of self-praise
- **Scarlet Fever:** too much use of color
- **Neglect:** many typos, poor layout
- **Constipation:** ideas not fully worked out
- **Malpractice:** insufficient experience, lack of qualifications
- **Vertigo:** a dizzying presentation. Too many ideas per paragraph. Too difficult to read.
- **Sleeping Sickness:** didn't get the attention of the reader. Boring.
- **Goiter:** swollen labor hours

<u>*Deliver and Follow-up*</u>

In this section, we discuss the proposal delivery, after the submittal, the debriefing, and the improvement process.

Let's talk about delivering a proposal. Sounds like a simple task, huh? Maybe not.

First, when should the proposal be delivered? The answer: one day ahead of the due date. Why? First impression, plus if you can get the client to give a quick review, you can fix any problems that come up. Also, you will be the only consultant who does this, and the client will not forget!

Second, how should the proposal be delivered? The answer: in person by the project manager is best! I wrote a proposal to conduct an environmental review of a mining site in Maine. We had a lot of competition. The client was located in Denver. I decided to fly to Denver with the proposal. The meeting went well, and the client was surprised that I made this extra effort. No other consultant traveled to the client. We won.

Third, be careful with overnight delivery! I've had two major proposals killed because the overnight delivery was delayed by a day. Always have a backup plan in place if you rely on a delivery service. By the way, another reason to deliver a day ahead!

Next, what happens immediately after the proposal is delivered? After you submit your proposal, selling goes into high gear! Always contact the client soon after submittal. There is nothing wrong with reaching out with a question like *did you receive the proposal?* You open a channel for communication that could be important. The client might ask you a question or tell you something important. Stay in contact on a regular basis. Follow the decision process closely. If you find an error in your proposal, fix it immediately. Be ready to make any tweaks suggested by the client.

The penultimate task is the proposal debriefing. Always get a debriefing, win or lose. *This is very important, and will impact your future success!*

Find out where your proposal was weak and where it stood above others. How was the PM received? How did the client judge the team as a whole? How about your technical approach or solution to the problem? Where did you stand with respect to schedule and cost? How did you rank against the other submittals?

What were the best parts of the proposal from the client's perspective? Can the client discuss each of the evaluation criteria with you? How did they see your price versus the *value…a critical question!*

Finally, one of your most important takeaways regarding a proposal is to learn from it, win or lose. Submitting a proposal creates an excellent opportunity to get valuable feedback on your work. You cannot afford to ignore this.

You cannot afford to keep making the same mistakes!

The project team should meet to discuss the debrief and suggest ways to improve. How did the effort compare with the nearest competitor? If you could resubmit, what would you change? What are the lessons for the future?

Keep good records for each proposal submission. Over time, your proposal win rate will improve!

Lastly, make every proposal better than the last one. Use a spreadsheet as a metric to track improvements and enforce this. Improving each proposal effort will virtually guarantee higher success.

Preparing Winning Presentations

In this section, we discuss how to create winning presentations. We will suggest a Success Equation, as we did in the previous section, followed by a few general pointers, a discussion of slide mechanics, some pointers on preparing the presentation, a suggested practice routine, and some delivery advice.

Presentation Success Equation

Does the title of this section sound familiar? Yes, it is the same equation we discussed in the proposal section. The terms are attention, value, educate, persuade, and anxiety.

Attention. You must find a way to get the attention of the audience, and have them listen *actively*. When the audience enters the presentation room, it is likely that they are thinking about other things. You must redirect their thoughts to the presentation by presenting something unique, provide an emotional message, or use a pattern interrupt (a surprise).

Here are five specific ways to get the audience's attention:

1. **Tell a compelling story.** This is a great way to open a presentation. People like stories. When you saw the last movie you loved, you didn't want it to end. A good story has a hero (that's you!), a back story (your history…your experience…accomplishments…success stories), enemies (in this case, your challenges that were overcome), disciples (your team members, or better, those who praise your exploits, such as a testimonial), and parables (your miracles or successes).
2. **Ask a situational question.** Start questions like the following: "If this happened, what would you do?"; "Have you ever…."; "How many times have you…."; Has this ever happened to you...." These questions make your talk immediately, personally, and emotionally relevant to each attendee.
3. **Use a catchy photo or graphic or a board**. This is a great way to get attention. When I am presenting to win a site selection study, I bring along a mounted detailed, colorful, GIS map.
4. Refer to **a recent, known current event.** Choose an event that attendees will be aware of and which creates an emotional response. In other words, an event they can identify with. For example, if one of their peers recently experienced a problem similar to the one at hand, the audience would relate to a mention of it. Talk about a recent assignment where you solved a similar problem.
5. **Use an aphorism or well-known quotation**. An aphorism is a familiar saying. Examples: "I prefer the errors of enthusiasm to the indifference of wisdom.", or "The secret to creativity is knowing how to hide your sources" (Albert Einstein), "There are two levers for moving men: interest and fear." (Napoleon), "Obstacles are those frightful things you see when you take your eyes off your goal." (Henry Ford)", "The proof of the pudding is in the eating." These are pattern interrupts.

Value. Value must be related to a business priority, such as making more profit, or minimizing risk or improving cash flow, or complying with new regulations. We covered value in the previous section on proposals. The same material applies here.

Educate. Talk about your understanding of the problem and the proposed solution, same as we discussed for proposal preparation. Remember, when you sell, you break rapport, but when you educate you build it. There is no resistance to educating the client.

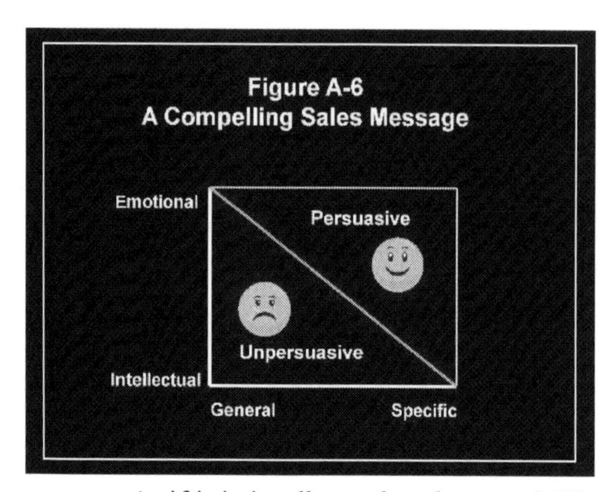

Persuade. In a presentation, a sales message must be persuasive. How do we accomplish this? A compelling sales message must be more emotional then intellectual or rational, and more specific than general or abstract, as depicted in Figure A-6.

A sales message will be *unpersuasive* if it is intellectual and general. The sales message will be *persuasive* if it is emotional and specific. By specific we mean include discriminators (in the form of benefits), a compelling metric, and a proven solution.

Refer back to Chapter 7 more suggestions about how to increase desire and aspiration, the backbones of persuasion.

Anxiety. Finally, use the same reasoning we presented for proposals. Remember to address both logical and emotional risk.

Some Presentation Pointers

Here are a few general pointers for your presentations:

- A great way to learn about how to present is to watch TED presentations. Notice how they begin and capture your attention. Look at their hand gestures and body movements. Notice the interaction with the audience. Observe the audience reactions. Note the closings as well.

- Is it a slideument? Is it a teleprompter? Is it a document? It is your presentation! If you are using slides in your presentation, they should present visionary ideas and concepts. They should support your communication, and not *be* the communication.
- What does your audience want? Once you know that, you must make an appeal to them. Clearly, you must know your audience and prepare accordingly. "What are they like?" This includes demographics and psychographics factors. These could include education, cultural background, interests, and personal goals. "Why are they here?" What do the prospects want out of the presentation? A solution? An approach? Empathy? "What keeps them up at night?" Is it fear of something happening? Pain? Risks? A thorn in their side? "How can you solve their problem?" What's in it for the audience? How are you going to make their lives better? "What do you want them to do?" State a clear action for the audience to take.

Let's finish this section with some typical problems we've all too often seen, that affect the success of a presentation. See Figure A-7.

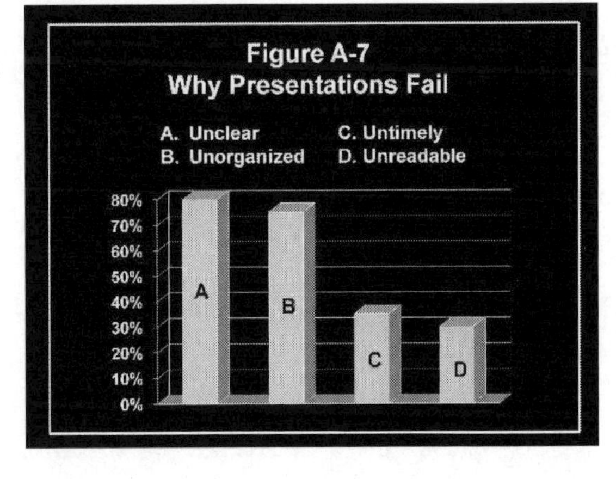

- The first is "Unclear". The presentation does not have a clear purpose. *It isn't focused on what the audience wants to hear*, or expects to hear. This is a common problem. It is easier to focus on what you know rather than what the client wants to hear.
- Next is "Unorganized", a made-up word to keep with words beginning with "u"! The *presentation isn't well organized and logically presented*. This is another very common mistake. It takes a specific effort to ensure that the presentation is clear and flows logically.
- Next is "Untimely". The presentation runs over on time, which either means you lose confidence by the listener (you didn't follow directions), or, worse, *you rush at the end and fail to bring your benefits*

home and give a clear call to action. This happens because of a lack of practice.
- Finally, we have "Unreadable". Your visuals are not easy to read, or too cluttered with data, or you rushed through them without providing sufficient time for absorption. Again, this occurs either because you didn't take time to view your slides in a room similar to where the presentation will be held, or because an objective reviewer didn't see a presentation practice session.

No excuses here. Keep these "uns" in mind when creating your presentation.

Slide Mechanics

If you use PowerPoint slides, or equivalent, in your presentations, these suggestions will help make your slides more effective:

- Use graphics, not lists, where possible. Images are retained better than words.
- Always use two channels, visuals and text, which enhances understanding and retention. What I mean by this is when you want to make an important point, use both text and an image to bring home the message.
- The first slide or two and the last slide or two are retained better than others. Prepare your sales messages accordingly. State your biggest idea at the beginning and repeat at the end.
- Break up your presentation with props, like boards or enlarged photos.
- Lists are really boring to read on a slide, and you will lose people's attention. Convert them to graphs if possible. For example, if there are no numbers, create a bar chart, and make the length of the bar equal to the priority or importance of the bullet.
- Ensure that the font is large enough to be readable from the back of the room. This is a common mistake. If the slides are difficult to read, you will lose your audience fast. Real fast. They will daydream, and listen to the voices in their head rather than your presentation!

- Use visuals in your presentations. A Wharton study says when trying to *persuade* an audience, visuals help a lot, as well as retention, comprehension, and attention. The study also indicates visuals improve clarity, interest, professionalism, and conciseness. Are you convinced yet that you should use visuals?

Preparing the Presentation

In this section, we will suggest a format for a presentation. I've used a presentation format, called a storyboard, or modular, format, since 1976, when I took a course on the STOP method of proposal and presentation creation. STOP was designed by Hughes Aircraft in the 1960's for developing complex proposals for government. Here's the definition of STOP:

> **STOP: Sequential Thematic Organization Process:** STOP works as an organizing tool that accommodates individual differences in presentation skills. The STOP method obliges the authors to search out and identify key themes before writing or presenting. Its use results in improved comprehensibility of the whole.

Storyboarding divides the presentation into parts, or segments. These segments are akin to chapters in a book. In this case, we typically use six parts. An objective is created for each segment. What are you trying to accomplish in each segment? Be clear about this. Once again, this is like a chapter in a book.

Storyboarding is essential for complex presentations; however, it can be useful for *any* presentation.

Storyboarding creates a flow that is easy to follow. It promotes logical thinking. Let's see how this works.

Typically, you can use one of three storyboard layouts:

- **Chronological.** If the topic has a natural timeline to it, use the chronological layout.
- **Problem to Solution**. If presenting the solution to a problem, use this one. I use this layout most of the time.

- **Logical**. If the topic doesn't match either of these layouts, just use a logical sequence.

A storyboard is comprised of a series of "panels", or "module's, or "chapters", as depicted in Figure A-8. Each panel may have supporting "sub-panels" or "subchapters".

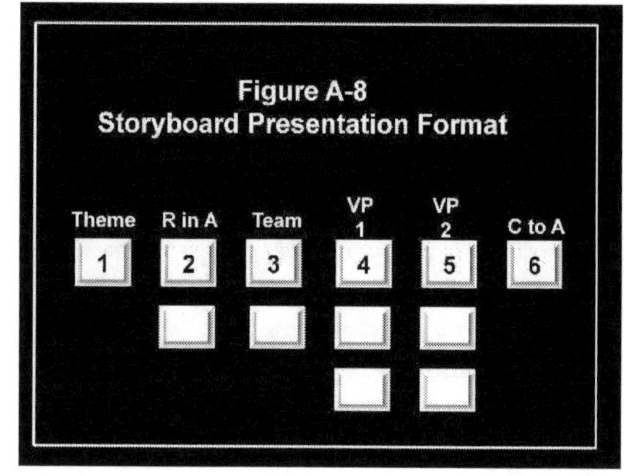

The panels are defined as follows:

- **Panel 1: Theme. This** is your compelling sales message, **the result you will deliver.**
- **Panel 2: Results in Advance (R in A).** Describe the solution to the problem, the major benefit, the key value provided.
- **Panel 3: Team.** Present the project manager and team.
- **Panel 4: Value Proposition 1 (VP 1).** What is your key value proposition? This is the first major benefit.
- **Panel 5: Value Proposition 2 (VP 2).** What is your secondary value proposition? This is the second major benefit.
- **Panel 6: Call to Action (C to A).** Ask for the job.

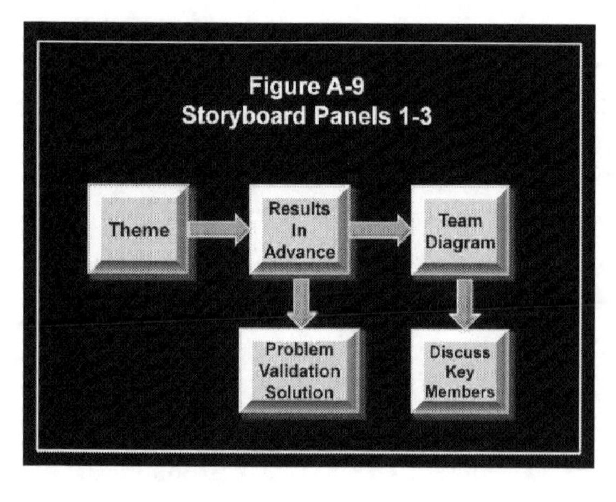

We will discuss each panel. The first three panels are shown in Figure A-9.

Panel 1, the theme slide, should state your compelling sales message or theme statement that includes the major or most important value provided (where a business goal the client wants is tied to a specific benefit generated by your offering), discriminators (benefits), a compelling metric, and a proven solution.

Here is an example from a previous project:

> *"Our approach to the City Solar Project will generate sufficient power to operate the Wastewater Treatment Facility and save the City $50,000/year, which can be applied to other important projects actively supported by the public."*

We call Panel 2 "Results in Advance" because you will show the client the results he will get by engaging you on this project. *Note that your theme statement may not convince your skeptical audience.* The Results in Advance slides begin with shining a light on the problem. Next, validate the problem using data (in other words, describe the challenge confronted by the client. Describe the negative consequences that will arise if the problem isn't dealt with now.). Finally, end with a description of the results, or outcome, desired by the client.

Here is an example from a previous project, and includes the problem, the validation, and solution with metrics:

> *"A town wants to recycle wine grape waste (pomace) into energy. Sending wine pomace to a landfill just fills the landfill cells prematurely."*
>
> *"The town could save 6 years of landfill life if pomace was treated."*
>
> *"Our biomass system will enable the town to recycle 80% of the pomace and generate enough energy to power the nearby wastewater treatment plant".*

In Panel 3, show the team organization chart. Use supporting slides to describe each *key* team member. Include their expertise (this adds credibility), role (for authority), and responsibility on the team (how they will help the client).

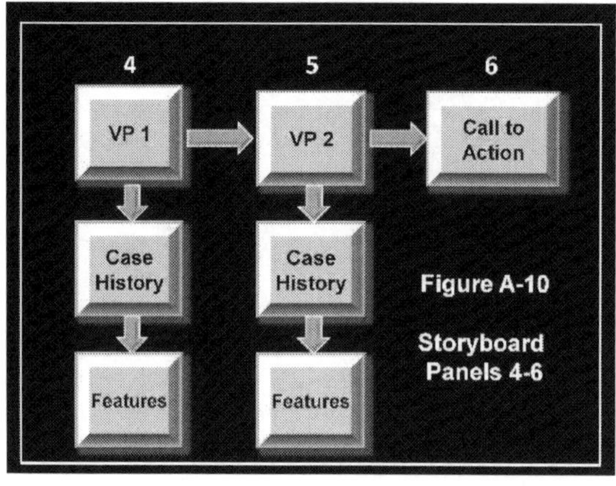

Figure A-10

Storyboard Panels 4-6

Panels 4-6 focus on the value proposition, as shown in Figure A-10.

Panel 4 presents the first Value Proposition slide. Start by explaining the biggest value, or benefit, you are offering to the client. Remember that value must be related to a business priority. Another way to express value is a Benefit – Cost, so address both the benefits gained (and more beneficial than your competitors) as well as the cost of the solution proposed. Remember that a value proposition also includes a comparison of alternatives. Then present a case history, where you in fact, delivered what you promised. Finally, support the value with appropriate advantages or features.

Here is an example of a Value Proposition panel (slide) from a recent project. It begins with a client benefit (actually a primary benefit followed by secondary benefits). Then a case history is presented (the York, PA project, operating since 2104). Finally, a few features are included.

The Albany project will:

Produce electricity for 100 homes in the City

Save 500 tons of carbon emissions per year

Demonstrate the City's commitment to renewable energy

This project mimics our York, PA project (2008)

<u>Project Features</u>

Uses the latest thin-film technology

Low visual impacts

$0.35/W below City's budget

Panel 5 is used for more complex projects, and will include another value proposition.

Panel 6 is the Call to Action Slide. What do you want the client to do at the end of your presentation?

The storyboard format has been a very successful approach to presentations.

Practicing the Presentation

"In theory, there is no difference between theory and practice. In practice, there is."
Yogi Berra

Well, I am sure Yogi knew what he was saying! Practice is essential for carrying out a presentation at the highest level of proficiency. You've got to practice. There is no other way. Timing of people presenting, the transitions, and the overall timing can only be firmed up through practice.

There are two types of practice necessary to hone the presentation:

- The proto-presentation is used for early dry runs. The purpose is to test the logic flow and messaging, or scripting. Don't worry about the details like transitions and timing. Interruptions are OK.
- The final dry run, which should be performed at least twice, should be used for refinements, timing, and transitions. Make sure you practice the final presentation against the clock, and aim for about two or three minutes short of the allowed time as a contingency.

Ensure that a couple people critique the presentation. Here are a few things the reviewers should watch for in the presentation:

- Are proofs (e.g. case histories) offered for the benefits?
- Does the speaker amplify the slide content, rather than just reading the slides?
- Are the slides too dense with words and numbers to the point that it discourages the reader?
- Can the slides be read easily at a distance?
- Are transitions between speakers smooth? Each transition should be the same.
- Do the speakers speak loudly and clearly?

- Do the speakers use any distracting movements?
- How convincing was the presentation? Is the talk persuasive?
- Does the presentation flow logically?
- Use the auto slide transition capability to check if your presentation can meet the time requirement.
- Did the first speaker introduce the team? My experience is that it is best for the first person to introduce the entire team who will speak. It is more difficult to have each speaker introduce the next one.
- Did the introduction establish the credibility of the key speaker? This is important. Not just role and responsibility, but why they are on the team!
- Does the speaker look at the audience? This is important for audience engagement.
- Do the speakers talk normally, not fast? Watch this carefully. Many speakers tend to start slow and speed up through their talk.
- Are there pattern interrupts in 5 to 10-minute intervals to keep attention?

Delivering the Presentation

In this final section of the chapter, we will discuss a number of issues around delivering the presentation, including room geometry, knowing the audience, delivering suggestions, and engagement hints.

The Presentation Room. Always try to see the presentation room prior to your talk. Check out the room geometry. Make sure your slides can be read from the back of the room. Where will you stand?

Try to get a small table so you can put your laptop on it and see your slide while facing the audience. This is important. You don't want to face the slides in back of you. That's a fast way to lose attention!

Make sure that you have your logistics in order. Every presenter knows where the room is. Everyone knows when they need to be there. Handouts are ready and available. Boards are ready and available. A laser pointer is on hand. If you are using the facility's computer, your presentation has been installed and tested.

Get familiar with the lighting and how to control it. Don't make the room dark. That is a fast way to lose an audience!

Delivering the Presentation. OK, you are ready to deliver the presentation! Get to the room ahead of time and get comfortable with the setup. Make sure the projection is working, files are loaded, working properly, etc.

Keep the lights on during the presentation, but not so bright as to make the slides difficult to read. You want the audience to see you!

Mingle with the audience, if possible, before the meeting starts. This will relax you, and help break the ice with the audience. NEVER apologize for a slide. Just acknowledge and correct an error and move forward. Distribute any handouts at the end. Stay on time. Don't talk too fast; it sounds like you want to get off the stage!

Tell the audience why you are different, not who you are. *Clients don't buy what you do as much as they buy why you do it.* You are there to persuade, not present! Remember to include persuasive language and techniques, not just rational statements, facts, and figures.

The audience doesn't care about *you*, they care about *them*. Don't talk about your expertise and your services as much as what benefits your prospects want from you. Create an upbeat mood before you begin.

Know Your Audience. Make sure you know the evaluation criteria, or explicit criteria. Can you put a priority or ranking, on these criteria? For example, cost savings, highly visible project, meet the timeline, reduced risk, match their needs well, public acceptability, increased safety, environmentally attractive, effective design.

Try to determine the implicit needs or wants, or personal benefits they are looking for. For example, security, recognition, no mistakes, high quality, very responsive to schedule, etc.

If there are several decision-makers, learn what *each* decision maker expects to hear in the presentation.

Look at the audience when you are speaking. This is important. When you look at the audience, and especially when you smile to the audience, it brings them to attention.

Keep the Audience Engaged. The biggest challenge in a presentation is keeping the audience engaged throughout. Here are six suggestions to keep your audience engaged:

- Stories are always a good way to interest the audience. Try telling a story to open your talk.
- Questions provoke thought and participation. Ask a rhetorical question they will think about. Ask open-ended questions. Start a topic with a question. A question engages the brain (that's how we are wired). Also, asking questions puts the presenter in control. For example, let's say you are presenting to win an assignment to repair a bad drainage problem on the High School ball field. Ask a question like "Do you remember when the Wildcats (the home team) had to run through 2 inches of standing water near the end zone last fall?" That's a good way to state the problem, and sets you up beautifully to ride in with the solution!
- Pre-frame the next slide. This provides a segue or transition between slides, raises excitement, builds curiosity, and creates anticipation. For example, "The next slide has the most important point I am going to show you today."
- Point out "new information". Everyone likes to hear about new stuff. Perhaps you can suggest a novel new technology to solve the high school drainage problem.
- Everyone *loves* new or innovative ideas. For proposals and presentations, innovative approaches likely are the very best way to engage a client.
- Use statements like "Is this clear/" or "Does that help" as pattern interrupts to bring the audience back to attention.
- Use the word "imagine. This word triggers the brain to instantly visualize almost anything. For example, ask the audience to imagine their future with your solution implemented.

Finally, avoid at all costs the dreaded Attention-Deficit Syndrome! This is what happens if you lose the audience's interest! You'll need a prescription from your presentation doctor!

For example, don't get into a string of slides without some type of pattern interrupt. Change things up. One thing you can do (we mentioned this earlier) is try breaking the presentation into "segments" of 5-10 minutes. Then, use a pattern interrupt about every ten minutes, at least one per segment. Examples are raising the energy level, asking a question, showing a prop, or transitioning to another speaker. Work to continuously improve your proposals and presentations, and you will soon be unstoppable!

Appendix B: Marketing Strategies for Your Key Clients

This Appendix addresses marketing strategies for your key, or most important, clients. Topics include sales strategies, client loyalty, Capture Plans, and strategic planning for key market sectors.

Sales Strategies by Client Type

All clients are not the same, in terms of how they should be marketed. There have been many attempts at "typecasting" clients. One approach is to break clients into three categories: loyal clients (a long-term source of business), one-off clients (these clients are more competitive, in that they spread work around to more consultants, and they are more difficult to keep over the long term), and low-price clients (they always select the low bid). Clearly, you would want to market the clients in these categories differently.

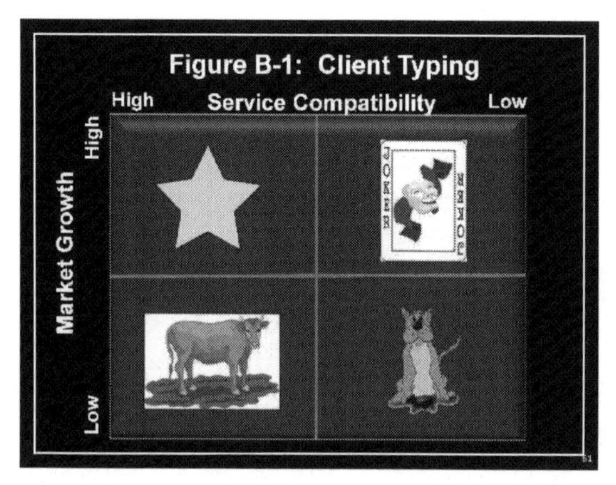

Another, more humorous example of client typing is shown in Figure B-1.

The plot shows service compatibility versus market growth. Service compatibility means how well your service matched client needs.

The yellow star represents client with high service compatibility and a growing market. The best of both worlds! The star of the bunch! The "cash cow" has strong market compatibility, but is in a slow market. So, milk it for all you can, while you can. The joker represents a growth market, but the service offering isn't very congruent with the client needs. You never know what you will get here. Perhaps you can get a few one-off opportunities. Finally, we have the dogs. Don't get bitten! This is a slow market with little service compatibility. Avoid!

We can define four types of clients: Key Clients (or Most Valuable Clients); Partners (or Master Clients); Price-based Clients (Commodity Services Buyer); and One-off Clients (Underperformers). Each of these is discussed below and shown in Figure B-2

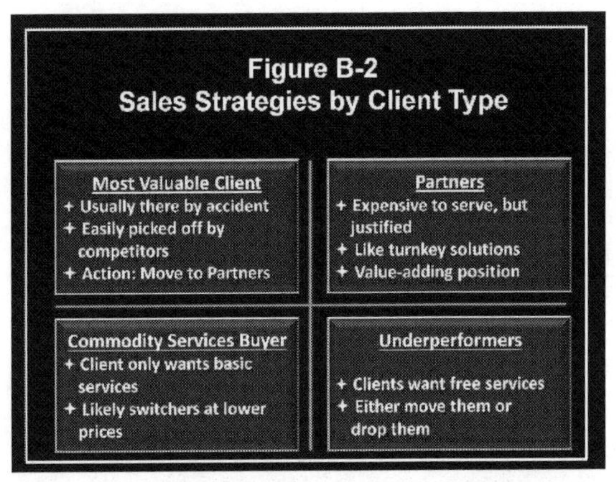

One-off Clients. These Clients are costlier to serve, and it is difficult to build relationships. These are the least valuable Clients. Obviously, *you want to expend minimal resources hunting or serving these clients.*

Underperformers will waste your time. You will write lots of proposals, and win few. They will drive your price too low. They will engage someone else on the next job. Drop them once they have been identified.

Price-based Clients. A commodity services_client isn't necessarily bad. These are the cash cows, although they do tend to spread the work around, and typically prefer the low-priced consultant. Price-sensitive clients are less costly to win and serve, and do not depend upon strong relationships. However, the competition will be fierce, and the winning price will be lower. *These clients should not be a priority unless you have a low-price strategy.*

Key Clients. Key Clients are characterized by strong relationships and relatively low serving cost. These are favored clients for sure. And, this is your stable of preferred clients from which you want to nurture Master Clients. Most Valuable Clients are good sources of business. However, you have yet to build the loyalty required to earn a long-term profitable client.

Master Clients. Master Clients are Key Clients that have moved to a *higher* level, that of an alliance partner or master contractor. Master Clients require higher serving costs, but it is worth it because you are earning more opportunities and revenues. Master clients under alliances tend not to duplicate expertise in their firms, but rather to support the alliance. These clients are very loyal, and should give you years of dependable work. Partners are costlier to serve, but the greater revenues justify the higher cost. Partners typically prefer turnkey solutions.

Type your clients and develop appropriate sales strategies for each client category.

Lifetime Net Worth

An issue that is important when marketing potential key clients, is the amount of effort and money we should expend to win a key client or expand work with a key client. I'm talking about the concept of "lifetime net worth", as described originally by Jay Abraham, one of the country's premier marketers. See, *Getting Everything You Can Out of All You've Got* (Abraham, 2000)

A lifetime net worth calculation can answer the question: how much should I spend on a proposal to win? If you've never used this technique, you might be surprised at how much time can be devoted to win a new client with an attractive lifetime value.

Let's look at an example calculation required to calculate the lifetime net worth. Follow these steps:

1. Estimate the lifetime sales for this prospect. Of course, "lifetime' is uncertain, so use the next 2 or 3 years.
2. Estimate the profit based upon this sales level.
3. Estimate the cost of acquiring the client/prospect.
4. The Marginal Net Worth (MNW) is (2)-(3).
5. Compare the MNW with the client acquire cost. If the MNW is larger than the acquire cost, you can increase the acquire cost.

Now let's look at some numbers. First, estimate the lifetime sales for the prospect or client. Let's assume that we have an estimate of $30,000/year for four years, or $120,000. The estimated profit based upon this sales level is $30,000, assuming a 15% profit.

Let's assume that the cost of acquiring the client is $5000. Therefore, the MNW is $25,000, which exceeds the acquire cost by $20,000. So, the numbers look good. Assume that you would need to spend up to $10,000 rather than $5000 to win this prospect. Should you do it? Sure!

What is the lesson from the Lifetime Net Worth calculation? Don't just base your proposal budget on the specific opportunity facing you. Consider a longer-term horizon. A larger commitment may be justified. Conversely, use this calculation to justify dropping a proposal opportunity because the MNW is negative.

Building Client Loyalty

Client loyalty is a critically important issue. Loyalty results in less volatile revenue streams and the ability to take some risks with new hires and innovative ideas. Build client loyalty one small step at a time and with consistency of performance. Here are nine ways to increase client loyalty:

- Keep learning. As you increase your skill set and knowledge base, you tend to share more with your client.
- Send thank your cards to your clients.
- Always observe birthdays and anniversaries. Ask the client's administrative assistant for the data.
- Invite your clients to events that you attend. Invitations are welcomed.
- Look for ways to educate your client about your service offerings, new technological developments, upcoming regulations, and market trends.
- Communicate often with your clients. Frequent conversations with your clients builds relationships.
- Ask for feedback. When you ask for feedback, it shows deep concern about your performance as it relates to them.
- Over-deliver. This is always an effective way to show a client you care about them.
- Ask your client how to improve. Go one step further than asking for feedback; ask how you can improve your service.

Susan Fiske and Amy Cuddy have performed considerable research into how people react to expressions of warmth and competence. Figure B-3 shows how a client perceives your behavior, as a function of competence and warmth.

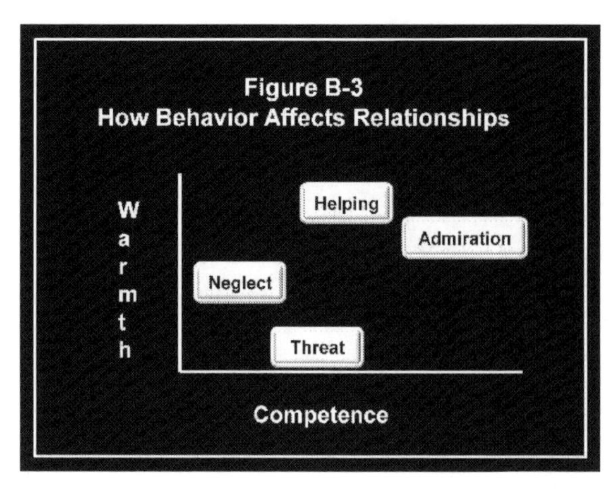

If you display low competence and low to moderate warmth, clients will want to avoid you. If you display low competence and low warmth, clients will perceive you as a threat to them, in the sense that your solutions will be detrimental to their goals. If you display moderate to high competence and high warmth, clients will see you as helping them with their problems. If you display high competence and high warmth, people will admire you.

Bottom line, you need to display both competence and warmth to succeed with clients.

A similar analysis centers on likeability and ability. When you meet someone, you will judge them by how likeable they are and whether they are competent, or have ability to do what they say they will do.

The likeability dimension includes friendliness, helpfulness, trustworthiness, and warmth. The ability dimension includes knowledge, skills, creativity, and productiveness.

These relationship dimensions make up 82% of the variance in the perception of social behaviors. Likeability (60%) is more important than ability (40%) because intent to harm is a survival instinct. In other words, we tend to assess a person's intentions before determining their ability to carry out their intentions.

Will the person harm me or help me? That is the first question. Then evaluate the ability of the person.

In summary, when dealing with clients, consider the issues of warmth/likeability, competency/ability.

Measuring Client Relationship Depth

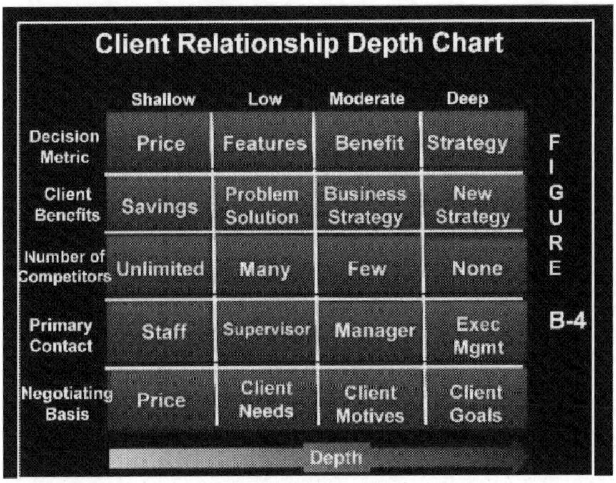

Figure B-4 shows what we call a Client Relationship Depth Chart. The purpose of this chart is to provide an indication of how deep your relationship is with your client, and provide guidance with regard to where you should focus your relationship building efforts. This chart isn't perfect, but I believe it can be *useful.*

Client Relationship Depth Chart

	Shallow	Low	Moderate	Deep
Decision Metric	Price	Features	Benefit	Strategy
Client Benefits	Savings	Problem Solution	Business Strategy	New Strategy
Number of Competitors	Unlimited	Many	Few	None
Primary Contact	Staff	Supervisor	Manager	Exec Mgmt
Negotiating Basis	Price	Client Needs	Client Motives	Client Goals

Depth →

FIGURE B-4

The vertical axis shows five relationship indicators: Negotiating basis; primary contact; number of competitors; client benefits; and decision metric. The horizontal axis is arranged in terms of a relationship depth indicator from "shallow", to "low", to "moderate", to "deep".

Let's look at one indicator: negotiating basis. Negotiating basis is broken into four levels: Price, Client Needs, Client Motives, and Client Goals. Negotiating on Price is at the shallow end of the depth scale, while negotiating goals is at the deepest relationship level. Clearly, if you are negotiating only on price, you are not in a good place with the client. If you are discussing needs with the client, you are in a much better place. Discussing the client's goals will put you in an advantageous position with a client.

Next, let's consider Primary Contact. If your primary contact level is a staff person, you are in a shallow relationship depth, one that does not provide much leverage. If your primary contact is at the executive management level, you have a deep relationship that will bring significant benefits to your firm.

Third, consider Number of Competitors. If you have many competitors for business at a client firm, you are at the shallow end, a place where your ability to win a competition is limited. If you have no competitors, you are at the deepest relationship level that will be envious to your competition.

Next, if your primary benefit is saving the client money, you are in the shallow end of the relationship. Solving problems is better. Working to help implement the client's business strategy is very good, but helping the client develop a new strategy puts you at the deepest level.

Finally, if the client hires you based on price, you are in the shallow end. Features aren't much better. If the client engages you because of the benefits you offer him, great. If the client engages you because you offered them a new strategy, you are in the cat bird's seat!

Use this relationship depth chart to measure where you are in your client relationship. The chart will tell you what you need to do to strengthen or deepen the relationship. The ultimate goal is to be on the right side of the chart on all indicators.

Pick one of your clients. Where are you on the chart?

Finally, look at the relationship from your perspective. What is your role with your client? Are you a contractor? A contractor provides a basic service to the client. *You are essentially acting as another staff member.*

A problem solver delivers a solution that the client needs. A project manager leads a team that delivers a desired result for the client. *You are in more of a leadership position.*

As a trusted advisor, *the client comes to you with their toughest, or most important*, or emergency, assignments. As an innovator, *you are viewed as a creator, a change agent*, who can deliver outstanding results that impact the firm's performance.

Where are you on this scale with your clients? I hope these simple tools help you see where you are in your client relationships, and provide you with some goals for the future.

Key Client Programs

A key client program is a systematic effort to build client relationships with clients deemed critical to your firm's success. The focus is on relationships rather than sales. Key clients are too important NOT to be managed by a system. You have a choice to view an important client relationship as a series of unconnected assignments, or a long-term, steadily expanding relationship that benefits both you and your client organizations.

There are additional benefits of key client programs, including:

- You want to *protect* your client relationship from competitors. The best way to do this is to put them into a special program, where they see a real commitment by your firm.
- You want to *maintain* this relationship and the benefits of the relationship to your firm. You've made investments in this relationship. You want those investments to create a return.
- You want to further *develop* this client. In a key client program, you can more easily add more contacts in the firm, and work both horizontally and vertically.
- Finally, a Key Client Program keeps your *focus* on the client. *This is likely the most important advantage!*

What does a Key Client Program look like? Here is one outline:

1) **Client History.** Describe the work that has been completed for the client to date, including the scope, budget and schedule compliance, primary client contact, your project manager, and your team. List the billing history as well.

2) **Client Goals.** What are the client's plans for the next year? What are the anticipated opportunities for your firm? What revenues are expected going forward?

3) **Executive Contacts.** Which of the firm's senior personnel have a relationship with the client?

4) **Service Delivery team.** Who is the program manager for this client? Who are the project managers who typically work for this client? Which staff is mostly responsible for service delivery to this client?

Key Client Programs are all about a team approach rather than a primary contact approach.

5) **Pricing Strategy.** What is your pricing strategy for this client? Do you offer a volume discount or other incentive?

6) **Success Criteria.** If there are no targets, the program will fail. How will success be defined? Will you use revenues? Will you use number of new projects? Will you measure the level of penetration of the client organization (horizontally and vertically)? Will you measure client satisfaction? Will you monitor the depth of relationships, and movement towards true partners?

7) **Meetings.** We suggest that three types of meetings be held with your key clients:

    1) Standard project management meetings, where you conduct project business. Use these meetings to identify future opportunities, and educate your clients about new services or other developments;

    2) Client feedback meetings, with the sole intent to determine how you are doing and how you can improve your services to the client; and

    3) Client appreciation meetings, held for the purpose of demonstrating the importance of this relationship to your firm. Express heartfelt appreciation for the trust they have in you. These meetings can be with a more senior member of your firm, or with members of the project team. *No business is discussed.* Take a client to play golf or watch a ballgame. Take the client and spouse to a nice dinner! We do not recommend that you combine the topics of these meetings into one meeting!

8) **Capture Plans.** The purpose of client capture planning is to put a strong focus on your key clients. A typical client capture plan includes the following:

1) Current status of work with the client;

2) Primary client contacts;

3) Your revenue and new project goals for the next year;

4) The client's plans for the next period of time;

5) An action plan that lists specific actions that will take place in the next month and quarter; and

6) Results of quarterly reviews of the action plan.

Strategic Plans for Key Market Sectors

In the previous section, we focused on a client organization. Now we move up a level and discuss the market environment that your key clients work in. We call this topic "strategic planning for key clients" or "strategic plans for key market sectors". Strategic client planning is a more detailed and holistic approach to planning for your basket of key clients.

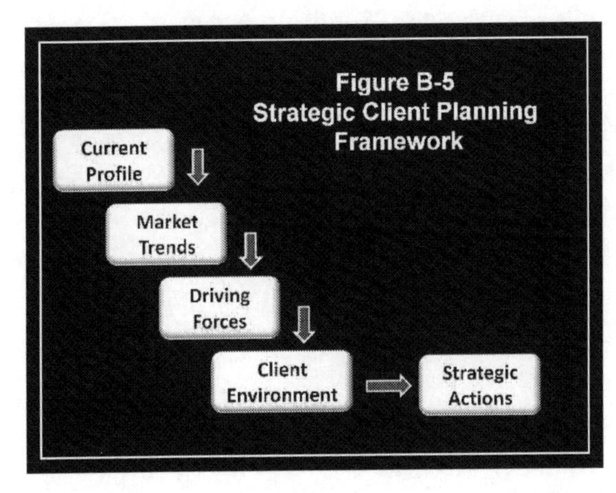

Figure B-5 shows a Strategic Client Planning Framework. This is a process I've used to conduct annual evaluations of market sectors where our key clients are located. The most critical part of this process is to look at the future, predict events, and make investments, despite the uncertainties. Real growth comes from leading the market, not following. This is a key distinction.

Each of the components in this framework is discussed below.

Current Profile. The current profile is a description of *where we are now*. The first step is to think about where you want to be in a year with regard to your key clients. For example, you may want to increase revenues by 20%, and add three key clients.

Next, summarize the progress you made in the past year. What worked, what wins, and why were there failures (e.g. proposal losses, job losses). Then list your core competencies. What are you best at doing now? In what areas are you positioned as the expert?

Then move on to your competition. Who are your toughest competitors? What are their core competencies? Who are their clients? What are they doing right? What do you need to compete successfully?

Finally, take a look at your strengths and weaknesses. The more you know about your firm's strengths and weaknesses, the better able you are to see opportunities. How are you doing in marketing? How about sales and proposal preparation? How about project management? Are you taking good care of your key clients? How well do your core competencies line up with market opportunities? Where has your market share increased? Decreased? List the top three weaknesses in staff and results. Where are you vulnerable, in terms of staff strength, materials/equipment, budget, etc. In summary, take a good look at where you are today. It is sort of an inventory.

Market Trends. What is trending in the market? Trends offer insights into where the opportunities will be in the next time period. First, look at your client's market sectors. How is the market doing? What market forces are at play? How will these trends affect your clients?

Second, what opportunities do your clients have in their marketplace? How are regulations affecting their business? Are there new regulations coming that will impact their business? Do you see any opportunities for you? Look at both the Federal and state level. What actions will be taken by local or regional agencies and utilities as a result of new or anticipated regulations? For example, I have a solar development business. In June 2014, the Obama Administration announced the Climate Action Plan. The goal of this plan is to reduce greenhouse gas emissions from fossil power plants. The EPA offered affected utilities various alternatives to reach their reduction goals. One of those alternatives is more solar energy development. As a result, we began to monitor utilities in good solar insolation areas to see who will select this alternative. Since then, even though the Trump Administration will not pursue these regulations, utilities are moving briskly toward more solar development.

Third, what threats are your clients facing? How is the current and anticipated regulatory environment affecting them? Can you help your clients develop risk profiles of new regulations?

Fourth, how are competitors in your firm's markets behaving? Look at your firm's competitors. What are they doing? What initiatives are they pursuing? What actions are they taking that you can copy, or offer a better solution? Why are they taking these actions?

Driving Forces. What is driving the markets your key clients are operating in? How is the economy affecting your business and your client's business? If the economy is getting more difficult, how can you help clients cope? If the economy is improving, what new services can be proposed?

How about environmental regulations? Are there new regulations coming that will affect your clients? Have unexpected changes in the weather created any opportunities? Are there any other regulations, or political actions, that will drive work opportunities? How about government incentives? For example, the solar energy market is very dependent upon government incentives, both state (Renewable Performance Standards) and Federal (construction cost rebates).

How is technology driving your business and your client's business? Look at social media and the Internet for example. How are they interacting with your business?

Are there any demographic changes that offer opportunities? Is there any new construction that will provide work opportunities? Look at each age group, including retirees.

Client Environment. Next, look for ways to gather information about your clients that will affect their businesses in the next year. There are lots of external sources of this information, such as annual reports, 10Q reports, and their websites. For example, do they have any new construction projects in the works? Is there any hint of upcoming mergers or acquisitions? What regulations affect their business? What are their key business risks? Also, being familiar with these issues will make you sound more informed. *Of course, the best source is to ask them!*

Is there spending increasing or decreasing? Why? What new markets are being considered by your clients? (e.g. solar energy for electric utilities, fracking wells for the natural gas industry, online learning by universities). And, of course, what needs will your clients have in the next year that *you* can address?

Strategic Actions. All of the preparation from the previous steps in the Strategic Client Planning framework must lead to actions. Note that your Strategic Plan will center on your key clients -- perhaps two or three clients in all -- in addition to a plan to attract one or two more key clients in the next year.

Identify the critical issues or insights that resulted from the strategic assessment process (current profile, market trends, driving forces, etc). These issues will bridge the gap between the current situation and the expected outcome for the following period.

Begin with new client development. This includes moving a commodity client to a key client as well as winning new clients who have the potential to become key clients in your existing market segments, where you have leverage. Also, target new prospects, as well as other divisions or departments of existing clients organizations. How will you convert some existing clients to key clients? Can you negotiate master contracts or alliance contracts? Formulate your "new client plan".

Next, what new staff expertise will be needed in the next year to address new client opportunities? How will you fill that gap – a new hire or training existing staff? Develop a New Hire Plan.

What new service offerings should be made available to clients? How will you provide those offerings: in-house development or outsourced?

What new markets or market sectors or market segments will you pursue? Why do you think this new sector is so promising? Do you have the required skills for this market? Do you have good intelligence about this market? Who will head up the marketing effort?

Do you need to make any changes to your organizational structure this year? Will changing the organizational structure help improve results in the next year?

Should the compensation profile be changed? Should certain staff be promoted with greater responsibilities for expanding your key client work?

In summary, performing the steps in the Strategic Client Planning framework should position you for a more successful year. Craft this strategy to do just that. Be sure that your actions will enable you to achieve your vision for the next year.

Operational Plan. Let's not forget this step. The easy part is to develop a plan. The hard part is to implement the plan. Who will be responsible for implementation of the plan? Senior management must support implementation. Implementation must be monitored via tracking and periodic progress meetings. What metrics will be used to indicate success? Make mid-term adjustments where necessary. Plans are plans that can be modified whenever appropriate. Your strategic client plans must be "living" plans.

The Think 2X Strategy

OK, you just evaluated your current profile, considered market trends, identified driving forces, investigated your client environment, and developed some strategic actions. There is a reason we brought you through this process. Your strategic actions have goals. For example, increase revenues from key clients by 10% next year. Rather, consider what Figure B-6 says!

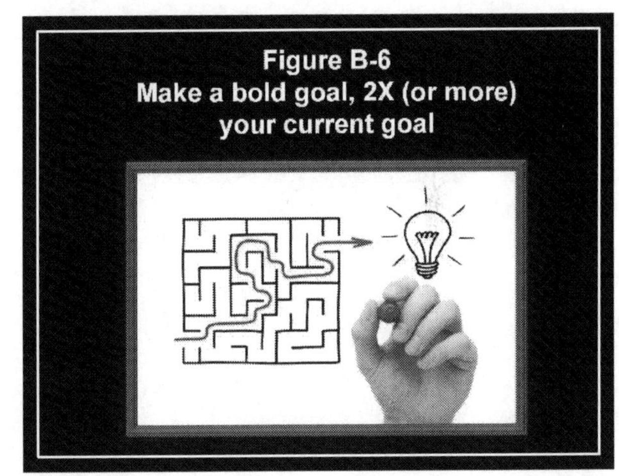
Figure B-6
Make a bold goal, 2X (or more) your current goal

Now consider a 100% increase in that revenue goal, what I call the 2X strategy. Set your goal at 20% increase in revenues from your key clients. The idea here is to change your mindset. Your eyes and ears only see and hear what the mind is looking for. Let's change that mindset to 2X.

What happens to your thinking when you think 2X? It has to change. By a lot.

2X will force out-of-normal thinking.

2X will make you more creative and more of a risk-taker. You will need this to reach 2X!

2X will increase teamwork, and stimulate resourcefulness if you want to reach 2X!

2X will energize your best staff, who are growth oriented, and love a challenge.

The 2X strategy will make you ask yourself questions like these:

- Is this process going to deliver 2X results?
- How must we expend effort to go to 2X?
- What resources do we need to achieve 2X?
- What processes should be eliminated because they cannot help you deliver 2X results?
- What new processes should be instituted to attain 2X?

The 2X strategy will force you outside the box. Your thinking will change. You will notice the difference. You will earn revenues beyond your current goals!

Go 2X!

Works Cited

Abraham, Jay. *Getting Everything You Can out of All You've Got: What to Do When Times Are Tough*. London: Piatkus, 2000.

Ackoff, Russell Lincoln. *Ackoff's Best: His Classic Writings on Management*. New York: John Wiley, 2010.

Amabile, Teresa, and Steven Kramer. *The Progress Principle Using Small Wins to Ignite Joy, Engagement, and Creativity at Work*. Boston, MA: Harvard Business Review, 2011.

Baker, Edward Martin. *The Symphony of Profound Knowledge W. Edwards Deming's Score for Leading, Performing, and Living in Concert*. Iuniverse, 2016.

Bowden, Adrian R., Malcolm R. Lane, and Julia H. Martin. *Triple Bottom Line Risk Management: Enhancing Profit, Environmental Performance, and Community Benefits*. New York: Wiley, 2001.

Burchard, Brendon. *High Performance Habits: How Extraordinary People Become That Way*. Carlsbad: Hay House, 2017.

Chamorro-Premuzic, Tomas. *The Talent Delusion: The New Psychology of Human Potential*. London: Piatkus, 2017.

Cialdini, Robert B. *Influence: The Psychology of Persuasion*. New York, NY: Collins, 2006.

Covey, Stephen R. *Quotes and Quips: The Seven Habits of Highly Effective People*. Provo, UT: Covey Leadership Center, 1993.

Dweck, Carol S. *Mindset: The New Psychology of Success*. New York: Ballantine, 2008.

Evans, James R., and David Louis. Olson. *Introduction to Simulation and Risk Analysis.* Upper Saddle River, NJ: Prentice Hall, 1998.

Hess, Edward D. *Learn or Die: Using Science to Build a Leading-edge Learning Organization.* New York: Columbia Business School, 2014.

Hill, Napoleon. *The Law of Success in Sixteen Lessons, Teaching, for the First Time in the History of the World the True Philosophy upon Which All Personal Success Is Built.* N.p.: ORNE, 2004.

Hogshead, Sally. *How the World Sees You: Discover Your Highest Value through the Science of Fascination.* New York, NY: HarperCollins, 2014.

Holmes, Chet. *Ultimate Sales Machine: Turbocharge Your Business with Relentless Focus on 12 Key Strategies.* Place of Publication Not Identified: Portfolio Penguin, 2015.

Kahneman, Daniel. *Thinking, Fast and Slow.* New York: Farrar, Straus and Giroux, 2015.

Keeney, Ralph L. *Siting Energy Facilities.* New York: Academic, 1980.

Keeney, Ralph L. *Value-Focused Thinking: A Path to Creative Decision-making.* Cambridge: Harvard U, 1998.

Kennedy, Dan S., and Kim Walsh-Phillips. *No B.S. Guide to Direct Response Social Media Marketing.* Irvine, CA: Entrepreneur, 2015.

Kirkwood, Craig W. *Strategic Decision Making: Multi-objective Decision Analysis with Spreadsheets.* Belmont, Calif.: Duxbury, 1997.

Mankins, Michael C., and Eric Garton. *Time, Talent, Energy Overcome Organizational Drag and Unleash Your Team's Productive Power.* Boston, MA: Harvard Business Review, 2017.

Pink, Daniel H. *Drive: The Surprising Truth about What Motivates Us*. New York: Riverhead, 2012.

Seelig, Tina Lynn. *InGenius: A Crash Course on Creativity*. New York: Harper One, 2015.

Seelig, Tina. *Insight Out: Get Ideas out of Your Head and into the World*. New York, NY: Harper One, 2015.

Sudan, Rafaella, Nicholas Bloom, and John Van Rennen. "Why Do We Undervalue Competent Management?" *MTP.*, 08 Sept. 2017. Web. 17 Oct. 2017.

Sugarman, Joseph. *Triggers: 30 Sales Tools You Can Use to Control the Mind of Your Prospect, to Motivate, Influence and Persuade*. Las Vegas, NV: DelStar, 1999.

Williams, Tim. *Positioning for Professionals: How Professional Knowledge Firms Can Differentiate Their Way to Success*. Hoboken, NJ: Wiley, 2010.

Made in the USA
Columbia, SC
27 November 2017